AREA ARRAY PACKAGE DESIGN

AREA ARRAY PACKAGE DESIGN

Techniques in High-Density Electronics

Ken Gilleo, Ph.D. Editor

McGRAW-HILL

New York Chicago San Francisco Lisbon London
Madrid Mexico City Milan New Delhi San Juan
Seoul Singapore Sydney Toronto

The McGraw-Hill Companies

Cataloging-in-Publication Data is on file with the Library of Congress

1 2 3 4 5 6 7 8 9 0 DOC/DOC 0 9 8 7 6 5 4 3

ISBN 0-07-142827-5

The sponsoring editor for this book was Stephen S. Chapman, the editing supervisor was Stephen M. Smith, and the production supervisor was Sherri Souffrance. It was set in Times Roman by Wayne A. Palmer of McGraw-Hill Professional's Hightstown, N.J., composition unit. The art director for the cover was Margaret Webster-Shapiro.

Printed and bound by RR Donnelley.

Portions of this book were previously published in *Area Array Packaging Handbook,* © 2002.

CONTENTS

Chapter 2. Trends/Drivers in the Electronics Manufacturing Industry 21

Chapter 3. Area Array Packaging 47

Chapter 6. Flip Chip Technology 103

Chapter 7. Options in High-Density Part Cleaning 141

PREFACE

Area array packaging provides the highest density with maximum performance of any concept yet devised, making it the most important technology in advanced packaging. Older and more familiar perimeter connection–style packages that first appeared about 40 years ago still serve many applications, but designs exploiting package area for connections are now essential for modern electronics, especially high-performance systems and portable products. This book begins with definitions of packaging, next details the numerous requirements, and later describes how packaging has evolved to satisfy these needs. Diagrams show details of various package constructions to highlight the all-important role as the "bridge" between semiconductor devices and circuit boards. Perimeter-style designs are compared to area array concepts with convincing arguments for the continued advance and deployment of this newer technology.

The book provides a comprehensive overview of chip- and board-level interconnection methods. Chip-level, or first-level, connections, such as wire bonding, TAB (tape automated bonding), and flip chip, are covered in detail. Increasingly popular flexible circuit–based packages, now extensively used, are described, showing how they utilize all three types of first-level connections. Substrate and routing considerations are discussed, followed by chip protection schemes. Finally, the package-to-board, or second-level, connection options are compared. Eutectic solder balls are the most common and are covered thoroughly, but high-melting columns, nonfusing metals, and polymer conductive adhesive options are included and contrasted.

Packaging materials are summarized to provide the necessary background to better understand concepts and designs. Package platforms, chip carriers, and substrates bring the focus to ceramics, metal, organic materials (including flexible circuits), and combinations. Materials topics include substrates, joining media, conductors, adhesives, encapsulants, fluxes, and underfills. Different solder alloys, including lead-free materials, are described, and the impact of resulting higher assembly temperatures is assessed.

The business overview chapter discusses needs, economics, and motivation for rapid and sustained growth for area array packaging. Both market drivers and technical benefit perspectives are included. A strong case is made for area array as the final packaging revolution since this architecture has retained key attributes from previous revolutions, like surface mount, but added new ones. The business and economic topics also show the importance of manufacturing technology, throughput, and yield on the electronics industry and the key role of packaging.

The book includes minimal area array designs like flip chip that achieve the smallest possible footprint, and stacked chip configurations that deliver the highest conceivable density. Undoubtedly, the most advanced designs will continue to employ area array concepts since a perimeter configuration is no longer an option for sophisticated active devices. Area array is essential for satisfying the mandate of "smaller-faster-cheaper," and new products continue to evolve. Modern cellular phones and virtually all personal computers now rely on area array packaging, and the role in these and other applications is described.

Since area array can provide the highest possible density, several chapters deal with designs and concepts that deliver the maximum, such as flip chip and chip scale packages (CSPs). Chapters cover details of flip chip, or direct chip attach (DCA), and discuss the subelements that include under-bump metallization, bumping processes, fluxes, underfills, and compatibility. Special attention is given to the underfill step, its present limitations, and future improvements.

CSPs are described from several aspects to provide guidelines for designers, assemblers, and users. Information in the CSP and flip chip chapters allows the reader to decide where each concept can best be applied and how to avoid the pitfalls of both. Advanced CSP and flip chip topics, including the latest multichip package (MCP) designs like 3D (stacked), are described in several chapters. Multichip structures encompass wire bonding, tape carrier packages (TCPs), and flip chip configurations for the first-level assembly within the package, as well as combinations. Applications are discussed showing how stacked area array packages, the densest package designs possible, are rapidly moving into computers and telecommunications products.

Reliability is covered from several viewpoints that include materials, design styles, and assembly. Comparisons are made between ceramic, plastic, and metal systems and protection strategies. Subjects include solder ball void defects and influence factors, flip chip bump types, underfills, and thermomechanical effects. The discussion of ceramic area array packages also compares reliability of ball grid arrays and column grid arrays (CGAs) and the benefit of greater standoff height. Reliability as a function of rework, higher-temperature solders, moisture adsorbed by packages, and assembly process effects is included.

This book also looks into the future to access the needs of the latest emerging semiconductor technologies and includes MEMS (microelectromechanical systems) and MOEMS, or optical MEMS. A new industry is unfolding based on chips that combine mechanics with electronics. The MEMS chapter points to the need for new packaging designs specifically for MEMS to satisfy mechanical device special requirements in this rapidly expanding and diversifying field. MEMS, and related optical chips, have special requirements, such as free space above the device to allow movement of the mechanical elements. Concepts for replacing costly metal and ceramic perimeter packages with MEMS-specific designs are suggested.

The reader should conclude that every important packaging topic has been included and that a convincing case has been made for the continued success and growth of the area array revolution far into the foreseeable future.

Ken Gilleo, Ph.D.

AREA ARRAY PACKAGE DESIGN

CHAPTER 1
INTRODUCTION TO ELECTRONIC PACKAGING

Ken Gilleo
Cookson Electronics

1.0 DEFINITION OF THE PACKAGE

The ideal electronic package is an economical and manufacturable electromechanical platform for one or more electronic devices that affords protection, facilitates handling, and provides the geometric translations and compatible interface required for connection to the next system level by practical assembly processes.

1.1 INTRODUCTION AND REQUIREMENTS

This book covers much of the field of packaging, including the new minimal category, but with a focus on area array designs. This chapter will describe the functions of the electronic package, protective requirements, and issues as well as solutions for general packaging. This chapter emphasizes area array packages because this is the central theme of this handbook.

The electronic package is designed to perform an assortment of tasks. Some functions can be at variance with one another, requiring balance and compromise for optimization. The package must first house and protect the delicate integrated electronics devices from the external environment. This generally is accomplished by enclosing the chip in a protective cocoon of electrically insulative materials. Alternatively, the package can be a metal vacuum-sealed container with appropriate internal insulation or a similar structure made of gas-impervious ceramic. The package also may need to dissipate substantial heat generated by the electronics. Heat management generally is not an issue for the hermetic metal package or the ceramic type. However, the common plastic package has good thermal insulation characteristics that prevent efficient heat transfer to the outside. Plastics have low thermal conductivity, so enhancing thermal performance of plastic packages takes ingenuity in both design and materials. The general electronics industry also seeks to reduce size and weight while the devices to be packaged become faster and have more connections. There is also strong pressure to accomplish this while lowering the cost. Thus the paradoxical phrase, "faster, smaller, cheaper." The packaging industry therefore is tasked with achieving goals that appear mutually exclusive but is asked to do so more quickly. Surprisingly, the industry has succeeded to a large extent, but solutions have required major innovations that are defined accurately as the "packaging revolutions" (Fig. 1.1).

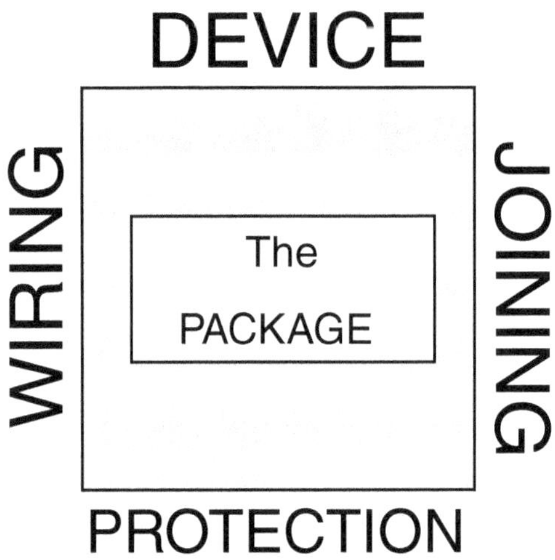

FIGURE 1.1 Packaging diagram.

1.2 PACKAGING EVOLUTIONS AND REVOLUTIONS

1.2.1 SMT

Most people are familiar with the standard molded-plastic packages seen in computers and their plug-in cards. Many of these packages are feed-through types where the metal leads fit through holes in the circuit board. The most common design in the feed-through class typically is called a *dual-in-line package* (DIP) because of the two rows of leads on each side of the plastic encasement. In the early 1980s, a more efficient packaging design strategy, called *surface-mount technology* (SMT), was introduced to solve a host of problems. The SMT process makes the connections simply by soldering the package leads to the surface of the board. The surface-mount device (SMD) can be much smaller, accommodate many more connections, is easier to assemble with automated equipment, and does not require holes through the circuit board. Absence of feed-through holes allows components to be assembled on both sides of the board. The DIP-type packages not only use up more surface area but also consume circuit density because holes must be fabricated through all layers of the board that otherwise could accommodate inner circuitry. The clear benefits of SMT have allowed this style package to capture more than half of designs. SMT has advanced over the years to reduce size and accommodate more leads. It should be noted that feed-through packages are still used, and this will continue. In fact, SMT and feed-through packaging can be complementary, and there are assembly processes for connecting both types in one operation. Figure 1.2 compares the feed-through style with SMT.

1.2.2 Perimeter Paralysis

In the early 1990s, it once again was time for a major change in packaging style. The relentless advancements in the semiconductor industry were putting an ever-increasing demand on even the densest SMD-style packages. The number of connections, or inputs/outputs (I/OS), moved continuously, while more compact electronic products demanded smaller packages. Since the packaging leads, or terminations, were on the perimeter, this meant that more leads were being placed in a smaller space. In other words, there was an ever-increasing "perimeter paralysis." Some of the packaging fabricators, most notably in Japan, responded by making smaller leads and placing them closer together for finer pitch. This approach soon ran out of steam as leads became flimsy, solder bridging defects increased, and assembly lines struggled with alignment problems. Ultrafine-pitch perimeter packages were the wrong solution.

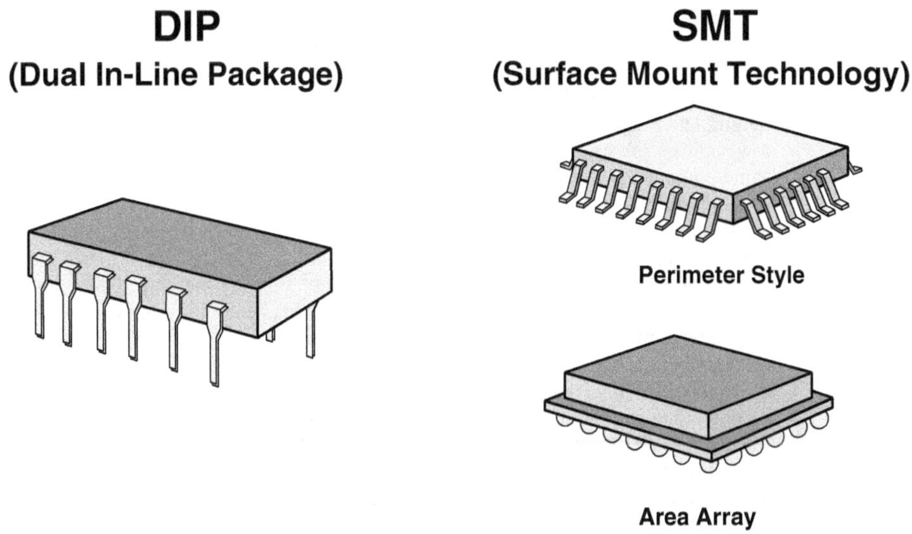

FIGURE 1.2 DIP versus SMT.

Vertically integrated electronics companies in the United States, led by IBM and Motorola, recognized that the time was right for the next packaging revolution. The simple answer for packing more connections into a shrinking housing was to use the wasted area inside the perimeter. Area array packages simply exploit the area under the package for connection sites. This was not an overlooked solution waiting to be conceived. IBM had used area array packaging since the 1960s. The issues were that a move to area array for ordinary packaging would be a hard sell and a daunting implementation task. IBM had developed ceramic area packages, with some using columns of high-melting solder, called *column grid array* (CGA). Motorola introduced a lower-cost *ball grid array* (BGA) in which solder spheres were used to make the connection to the board. Motorola's announcement of the Ompack BGA caused a major stir within the packaging and assembly industries. Many of us view Ompack as the official launch for the area array packaging revolution that is still vigorously underway. The plastic Ompack, or *o*ver*m*olded *pac*kage, generally is referred to as a *PBGA,* with the *P* standing for plastic. The PBGA is the most common type of area array electronic package. This book will deal with design, manufacturing, testing, and assembly of PBGAs and other area array packages as well as the materials.

There are many subdivisions of the BGA, but all are surface mountable, and most are area array. The next discussion will include the minimal micropackage beginning with the flip chip. While packaging is necessary, many view it as a "necessary evil" that should be minimized. Admittedly, the package tends to add cost and increase the size. Many have sought to reduce it to a minimum; hence the minimal package. I will avoid the clumsy terms *minimalist package* and *minimalistic package* to stay with the more appropriate and more minimal term *minimal package.*

1.2.3 Direct Connections

I will start with the simplest and most minimal package, the flip chip that was pioneered by IBM and referred to as C4 or C^4, for *controlled collapse chip connection.* Actually, it is necessary to take just one more step back to find the beginning of area array, SMT, and BGA—all in one package. In the early 1960s, while the industry was still working on transistor packaging, IBM came up with *solid logic technology* (SLT). This package, shown in Fig. 1.3, appears to be the very first minimal package: the first BGA, the first SMD, and the original flip chip. A more generic and inclusive term for this type of package is *direct chip attach* (DCA). Some people may argue that the SLT is not a chip, or integrated circuit (IC), but rather a lowly transistor. However, a transistor requires most of the same packaging attributes as an IC, and this design approach certainly could be used on an IC. The SLT has the essential constituents for the minimal package: protection and the electromechanical interface to the circuit board. The connection balls are made of copper and are soldered to the chip pads with a high-melting alloy. The nonfusible copper provides testability, an assured standoff height, and the potential for reworking the device. While most of today's BGA spheres are made of low-melting solder, copper spheres have been reintroduced recently by Alpha-Fry Technologies, the same company that made the original microspheres for IBM. Thus the newest area array packages may be reaching back almost four decades for "new" solutions (Fig. 1.3).

The now popular DCA, or flip chip, is the only system that directly attaches an IC to the printed wiring board (PWB) or to a chip carrier. The generic and descriptive term is *direct chip attach* because the important feature is the direct connection, not flipping the chip. Other packaging, such as *tape automated bonding* (TAB), can have a flipped chip; hence the confusion. Unfortunately, the more appealing name, *flip chip,* has become dominant and also will be used in this book. The DCA is most commonly attached to the substrate through small metal bumps that are formed on the chip pads by methods that will be described in detail in Chap. 6, "Flip Chip Technology." The most popular attachment method is to simply make the DCA bumps out of solder and reflow them to the substrate pads. The assembled component needs protection and mechanical enhancement, especially when bonded to a substrate with higher thermomechanical expansion. A polymer composite called *underfill* is used in most cases. The DCA assembly, once underfilled, has reasonably good reliability. The in situ "package" created with typical underfills cannot be removed in the practical sense. This means that the underfilled chip probably cannot be reworked, although special reworkable underfills recently have come to market. Some people would argue that a package must be removable and should be reworkable in the true and practical sense of the term. So is the DCA flip chip a package? Let's define the terms of packaging requirements and try to separate absolute essentials from desirables.

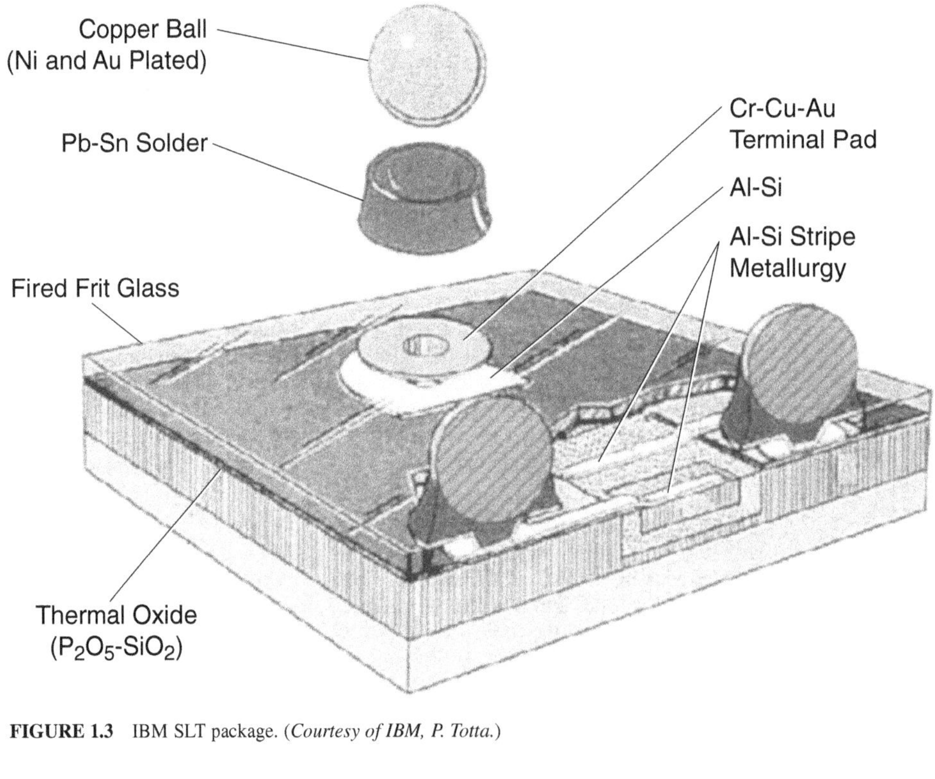

FIGURE 1.3 IBM SLT package. (*Courtesy of IBM, P. Totta.*)

1.3 USEFUL VERSUS ESSENTIAL FEATURES

The essential features of the package listed in Table 1.1 are required to make the semiconductor chip compatible with the printed circuit board (PCB and to give it reliability. The IC and PCB, or printed wiring board (PWB), used to avoid the other acronym associated with a pollutant) come from different worlds with different materials. The package bridges the inherent incompatibility between the microworld of inorganic ICs and the more macroscopic environment of organic PWBs. Not only does the most common package design expand and fan out the IC's microgeometry to fit the circuit board, it also allows the very thin aluminum (and now copper) chip pad to be connected to the relatively thick copper circuit pads. The list of useful features has become the main focus of modern packaging and will be discussed briefly here.

Table 1.1 is a list of all the both essential and useful attributes of the electronic package. In some applications, a useful attribute can be essential. Thermal management may not even be useful for a low-power memory chip but a critical requirement for a powerful central processing unit (CPU).

1.3.1 Geometric Translation

The IC device is made using processing that can produce features sized down to nearly 0.10 μm or 1×10^{-7} m, or about 1/10,000 in. Microelectronics is soon to become nanoelectronics as the feature size falls below 0.1 μm (100 nm). While the prefix *micro-* means one-millionth (of a meter), the prefix *nano-* means one-billionth (1×10^{-9} m). The printed circuit industry generally considers fine line to be about 3 to 4 mils (75 to 100 μm). There is nearly a 1000-fold feature size disparity between the two industries and their products. We can expect this ratio to remain somewhat constant or even to increase. The translator, the technology that spans the breach, is the package. The package has had to add a greater level of translation capability each time that the semiconductor industry has surged ahead with a smaller IC that required more connections or I/Os. The increasingly common strategy is to take the chip's perimeter pad connections and reroute them into an efficient area array. Although very high density interconnect (HDI) packages can accommodate chip pad dimensions, the HDI substrate is more expensive and only available from a limited number of suppliers. The economics can be better with a fan-out package, especially when many low-density components are used with one or a few high-density ones. Adding to the cost of packaging can save on the printed circuit costs to provide a net gain, especially if yield improves.

1.3.2 Material Compatibility: IC to PWB

There are two important material differences between the IC and PWB that make them incompatible. First, the IC substrate is inorganic silicon, with silicon compounds used as insulators. The most common PWB is made from epoxy resin with glass reinforcement. Silicon has a coefficient

TABLE 1.1 Essential and Useful Attributes of Electronic Packaging

Attribute	Absolutely essential	Very useful (not required)
Geometric translation	Usually	
Compatibility: IC to PWB	X	
Environmental protection	X	
Mechanical protection	Yes, most cases	
Handling ease		X
Enable standardization		X
Facilitate auto assembly		X
Removability	X	
Reworkability	?	X
Performance enhancement		X
Thermal management		In many cases

of thermal expansion (CTE) of just under 3 ppm/°C, whereas the CTE of the PWB can range from about 15 to more than 25 ppm/°C. This nearly 1:10 expansion difference can create considerable stress during temperature changes for a directly coupled chip. The package must intercede to alleviate or somehow compensate for this potentially destructive situation. The next incompatibility results from the differences in conductors. ICs typically use thin vacuum-deposited aluminum for conductor layers that results in a connectivity challenge. Even though some segments of the industry are changing to copper, its extreme thinness, a fraction of a micron, makes a direct connection difficult without preprocessing. First, these metals oxidize, and the oxidation must be removed for soldering, making the chip pad even thinner. The soldering process dissolves some of the substrate so that the thin copper or aluminum is dissolved, or leached away. In the case of aluminum, solderability is difficult at best even if the aluminum is thick. Something must be added to the pads. The package must make the chip pads compatible by adding additional conductive material, and adding more metal does this. However, the metal can be a coating, bumps, or wires, and this determines the basic connection class (covered later). Conductive adhesives cannot be used directly on IC pads because the native oxides are an electrical barrier.

1.3.3 Environmental and Mechanical Protection

All ICs are potentially sensitive to the external environment. Moisture, oxygen, and contaminants are usually harmful. Static electricity, even at low voltage, can blow apart sensitive microscopic elements within an IC. Moreover, mechanical handling can destroy many types of electrical chip connections such as wire bonds. Nearly all chips must be protected by encapsulation or by sealing within a preformed metal or ceramic hermetic housing. Even direct-attach and "chip-on-board" assemblies need protection, even though the "package" may seem to be absent. For example, a DCA assembly typically will use underfill within the gap between the chip and PWB to control strain. The underfill also serves as a protective encapsulant that prevents moisture and contaminants from reaching the active side of the chip. The underfill adds the extra strength needed for conductive adhesives and also reduces the potential for electromigration. Applying encapsulant after assembly can be called *postpackaging* or *in situ packaging.*

1.3.4 Handling Ease

The vast majority of the approximately 1 trillion chips produced each year are packaged using standard form factors that are handled easily by modern electronic component assembly lines. The common package is hard, strong, and easy to grip. The simple design allows easy recognition by visual inspection and by machine vision. The lead layout and package shape permit easy orientation. An SOT-23 transistor package, for example, has two leads on one side and a single lead on the opposite side. Not only can a pick-and-place robot easily confirm the orientation, but simple mechanical feeders also can orient this nonsymmetric package so that all are presented correctly. Packaging and machine-building associations have worked together to maximize compatibility between package style and machine, and this leads to the next subsection.

1.3.5 Standardization

Packaging standards bodies, such as the well-known Joint Electron Device Engineering Council (JEDEC), review proposals for new package designs and either provide suggestions for improvement or accept them into a standard. JEDEC is now the Solid State Technology Association (SSTA), a semiconductor engineering standardization body of the Electronic Industries Alliance (EIA), a trade association that represents all areas of the electronics industry. The approved standard is then published and distributed. Members and others agree to produce packages within the standard.

1.3.6 Facilitate Auto Assembly

The packaging and equipment makers work closely together to ensure compatibility. Several packaging standards organizations approve packages after sufficient time to comment. Assembly machine builders are represented by the Surface Mount Equipment Manufacturers Association (SMEMA) in the United States, which helps ensure that equipment meets standards that allow different brands to work together.

1.3.7 Removability/Reworkability

While most people agree that a package should be removable, not everyone insists that it be reworkable. *Removability* implies that the circuit board or substrate can be reused and that a new package can be assembled. Many people feel that *reworkability* also means that the package can be removed and then be placed and reconnected to substrate. This requires that the terminal leads be disconnected and then reconnected without damage or using a complex process. A removal process requiring chemicals and mechanical cleaning probably is not acceptable. The heating or other assembly processes should not degrade any of the structures during the removal and reattachment processes. Standards may require that the remove/add process can be repeated several times. The underfilled flip chip cannot be removed easily, and some people do not consider it to be a true package. Reworkable underfills use troublesome processes and even require mechanical cleaning steps and may not be reworkable in the accepted sense of the term.

1.3.8 Performance Enhancement

The standard package typically will degrade electrical performance, although only slightly for most applications. The connector length increase that results from the wire bonds, lead frame, or various other packaging connectors adds resistance, inductance, and time delay. The added resistance is insignificant for all but high-power applications. Inductance can add signal delay and distortion, but it has been a problem mostly at very high frequencies. Special packages are used for very high radiofrequency (rf) applications. However, as digital signal clock rates increase, the impedance characteristics (the combination of resistance, capacitance, and inductance) of the package will become more important. Packages that are designed to handle these higher frequencies can be said to enhance performance. Enhanced performance goes beyond simply protecting the chip. Some characteristic of the total system should be improved over the results in a more standard package design. A package substrate with a low dielectric constant k would enable better signal transmission at higher frequencies and would fall into the enhancement class. Underfill is considered packaging material with performance-enhancing qualities. The material can boost the thermal fatigue resistance of the assembly by a factor of 10 or more. Thermal transfer enhancement is such a critical area that it is a separate topic covered in the next subsection.

1.3.9 Thermal Management

All packages generate some heat because no electronic processes are 100 percent efficient. Their inefficiencies produce by-product heat. Some ICs can produce considerable heat, around 100 W for powerful new CPUs. The faster clock rates, greater number of transistors per chip, and diminishing size of the chip all come together to create a potential thermal catastrophe. The chip produces heat faster than it can be dissipated, causing the temperature to rise. The hot chip becomes even more inefficient and produces a higher output of heat. This thermal runaway condition finally reaches a level where the chip breaks down and is destroyed. A package must be designed to transfer this heat to the outside environment, either passively or with external coolers such as a fan.

The two thermal pathways are from the backside of the chip or through the metal chip pads on the front. A die, prior to wire bonding, is attached to a lead frame or chip carrier using die attachment adhesive that generally is filled with metal or mineral filler to increase thermal conductivity, as shown in Fig. 1.4. Some heat is removed through the connection pads, but this is minor when thin gold or aluminum wires are used. Tape automated bonding (TAB), covered in the next section, provides a better thermal path because more metal conductor with shorter paths is used. The best thermal conduit for the chip pads is with flip chip structures. The connection bump is very short and relatively thick. The flip chip package is one of the better thermal dissipation formats because heat also can be removed from the back of the chip that faces upward. Some flip chip packages, especially flex ball grid arrays (FBGAs), have a built-in metal heat spreader bonded to the flip chip

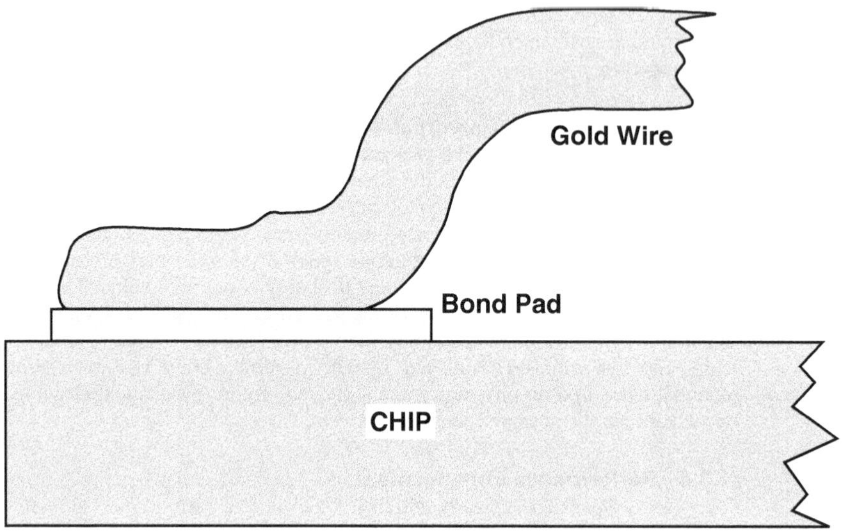

FIGURE 1.4 Thermal compression (T/C) bond with die attached.

that can be mated with a large heat sink for maximum heat management. New underfills will have increased thermal conductivity, thus adding an additional thermal path for flip chips.

1.4 EARLY PACKAGE DEVELOPMENT

Now that modern packaging principles have been discussed, let's review earlier packaging development. The need for packaging began with the first electronic device of any sophistication, and that, arguably, was the vacuum tube. The filaments, plates, and electrodes were protected in a glass vacuum tube that provided the mechanical and environmental protection while enabling easy assembly with a plug. The vacuum tube, which dates back to the twentieth century, provided all the essential features in Table 1.1 and many of the useful ones. Figure 1.5 shows an early Fleming triode in its very functional package. Vacuum tubes are still popular today for audio equipment and high-power transmitters. One can argue that the socketed vacuum tube is the first pin-grid array.

Modern packaging was launched with the solid-state electronics revolution, the first in electronics. The invention of the transistor in 1947 represented a major breakthrough that was not fully appreciated for many years. Transistor manufacturers quickly designed and implemented packages, and many are still in use today. Most were metal cylinders with three protruding wires, such as the still-popular transistor outline (TO) package. The eventual integration of the transistor to produce the IC would launch much more complex packages a decade later. However, probably IBM designed the most advanced and innovative early packages for the transistor in the early 1960s. IBM's solid logic technology (SLT) concept, described earlier, was far-reaching. This package has many of the features found in most modern micropackages. This direct-attach predecessor of the C4 flip chip is probably the first area array package ever used. Unlike most BGAs and flip chips, this package used copper balls instead of solder. Theoretically, the SLT package could be attached to substrate with eutectic solder and reworked because the balls are attached to the die with high-melting solder. Perhaps the SLT will be reborn.

Although the copper balls were attached with some level of automation, a mass bumping method was sought. Vacuum deposition was chosen because of the high degree of precision and excellent control of thickness. Depositing one metal and then another could form metal alloys. Since the metal

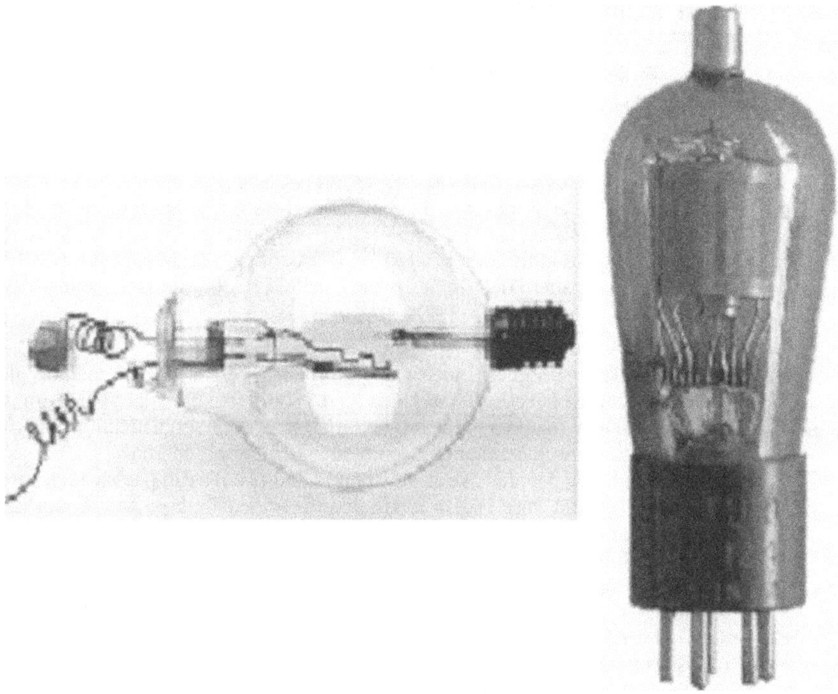

FIGURE 1.5 Fleming triode in a glass hermetic package (Prismark) and a glass vacuum tube.

columns were reflowed to form spheres, a homogeneous solder mixture resulted. These bumps allowed the chips to be soldered directly to substrate using controlled collapse chip connection (C4). C4 and flip chips are covered in detail in Chap. 6, "Flip Chip Technology."

Wire bonding (WB) was being developed independent of several DCA semiconductor producers. While IBM sought a very dense and extremely reliable main-frame computer package system, others needed efficiency, versatility, and low cost. The most important attribute of WB was the ability to "program" the wiring locations and thus eliminate tooling. WB also could be used on chips as received. WB allows the die size and pad configuration to be changed within reasonable limits without a package or lead-frame change. This feature is still extremely valuable today and accounts for the continuing popularity of the method.

Tape automated bonding (TAB) was developed shortly after WB and was sought as a means of simplifying the WB process. The idea was to fabricate a pattern of metal wires that could be bonded to the chip in a single gang-bond step. The original concept used metal with no dielectric carrier. A metal frame around the perimeter held the pattern of cantilevered wires. This frame was cut away after bonding to the chip. Now the "spider package" could be compression bonded or soldered to the circuit board. This bare-die concept evolved into an overmolded package. While several single-metal-layer packages have come along, various problems occurred, and the spiders became lead frames that were connected to the chip by WB. Not only did the wire bond allow different chips to be connected to one frame pattern, the wire served as a strain relief.

In the mid-1960s, true TAB, or flex-based, circuits were developed using conductors on a thin dielectric film. While the earliest idea used heat-sensitive DuPont Mylar (polyester), the products quickly moved to high-temperature Kapton (polyimide). Today's TAB or tape carrier packages or flex-based packages use polyimide. However, low-cost, low-lead-count polyester is used for calculators. The outer leads are connected using conductive adhesive film to avoid soldering.

1.5 ELECTRICAL CONNECTIONS

1.5.1 First Level: Chip to Carrier

The chip-to-carrier connection, generally referred to as *first level,* is a basic designed differentiation for area array packaging. There really are only two fundamental methods of connection to chip pads but several subcategories. The chip pad is connected either directly by means of a miniature bump or by a fine wire. Even TAB can be viewed not as a wire connection, but a pattern of wires that is chemically fabricated and held in a pattern. However, most authorities recognize three first-level connections: (1) DCA, (2) WB, and (3) TAB, or tape (flex-based). All three of these chip-level connections can be used for either full-sized BGAs or chip-scale packages(CSPs). The geometry of the package wiring pattern determines if the small CSP footprint is feasible.

1.5.2 DCA (Flip Chip)

DCA provides the highest electrical performance or speed. The method also allows the greatest number of chip connections and chips, and 7800 bump chips are being implemented. A disadvantage is that DCA requires precision and a high level of sophistication. We can view flip chip (FC) as a stand-alone package that is assembled *o*nto a circuit *b*oard (FCOB) or as a component *i*n *p*ackage (FCIP). Some people have argued that FCOB is not a full package because it cannot be removed once underfilled. A case is also made that the underfilling process is not part of the normal SMT assembly process and a great burden for the assembler. This may be somewhat valid, although new underfills are simplifying the process. The DCA high-density footprint also can challenge board-making capability. These issues have helped boost the chip-scale package.

However, there is another answer to these concerns: the FCIP. The FCIP is removable, reworkable, does not need to be underfilled by the assembler, and can alleviate some of the board density requirements. The resulting package typically is area array and can be designed easily as a small footprint fitting the CSP definition. The trend for all CPUs and many other flip chip components is toward FCIP.

1.5.3 Wire Bonded (WB)

While joints used in flip chip (DCA) provide a simple direct path, the alternative WB method from the late 1950s is much more popular. The amazingly simple but extremely versatile WB connection process has been the standard of the packaging industry for many decades. The method involves forming a reliable metallurgical bond between a very fine wire and a chip pad, looping the wire to the substrate, and forming the second bond. The versatility results from being able to cut the wire to a programmed length and to bond onto the desired location of the substrate. Very thin gold wire is used commonly, while aluminum is employed for simpler devices requiring thicker wire, such as power chips, but using an ultrasonic (U/S) process instead of thermal compression (T/C). The popular T/C process first forms a ball at the end of the wire using a hot flame such as hydrogen. The ball is pressed against the bond pad with heat and pressure, forming a metallurgical joint. The process is shown in Fig. 1.6, and it is used in the majority of area array and other packages for first-level assembly. The WB process can be used for just about every type of package, including micro-BGAs.

1.5.4 Tape Automated Bonding (TAB)

TAB is a flex-based package that has been around for many decades under many names. The most recent revival of this package class was by Intel, which chose to call it a *tape carrier package* (TCP), a more appropriate name. TAB, or TCP, is actually a specialized flexible circuit made with thin dielectric film with a "window" for bonding. The most common flex packaging material is high-temperature polyimide, which is able to handle processing temperatures well above 300°C. Flexible circuitry offers special benefits for packaging, including thinness, lightness, ultrafine pitch, and strain-relieving compliancy. A pattern of "wires" is formed from the copper that can be connected to the chip pads. An opening, called a *window,* is formed in the flex dielectric so that the metal conductors become overhanging cantilevered beams. Now the TAB can be placed over the chip and the beams aligned over the chip metal bonding pads. These metal beams, or fabricated wires, are then bonded using thermocompression bonding. Wire bonding and TAB inner lead bonding (ILB) have similarities. However, it is possible to gang bond the TAB leads, making this step

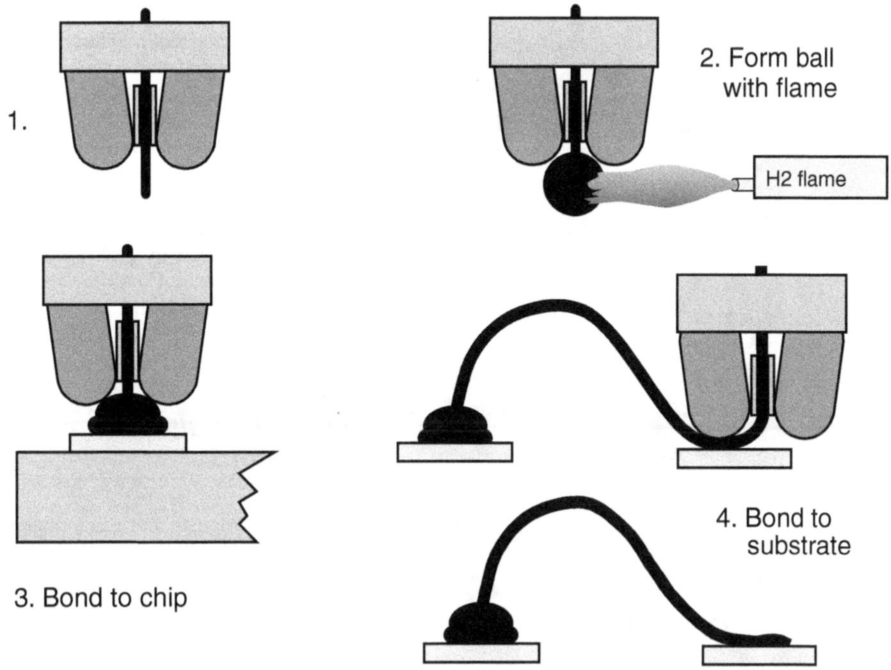

FIGURE 1.6 Thermal compression (T/C) bonding process.

very productive. The TAB inner leads typically are gold plated. The chip is often gold plated and may even have gold bumps to aid the connection process. The ILB process is efficient and reliable. The resulting package can be singulated from a reel of "tape" or kept in this form; hence the original tape name. Figure 1.7 shows TAB. The package generally is assembled to the circuit board by soldering the outer leads with a hot bar. The outer lead bonding (OLB) also may use anisotropic conductive adhesive (ACA) film. While the ILB process is a reasonably good one, the OLB step has to be the notoriously troublesome step. Later I will show how some of the newer area array packages have preserved ILB but eliminated the noisome OLB.

Density Ramifications. The DCA connection scheme provides the highest density because it is the only one that can use the full area of the chip. TAB can be problematic because it is impractical to connect to an area array chip. In principle, WB could be used to connect beyond just the perimeter of the chip, but the resulting "rats nest" would not be efficient, and signal crosstalk would be likely. However, WB or TAB over active die is considered potentially destructive. The mechanical force could fracture transistors and other devices or reduce their lifetime. The DCA bump can be produced by gentle nonmechanical processes that will not injure the delicate IC structure. The DCA bonding process is also gentle, making area array chip connection quite practical only by this method. The DCA will remain the highest-density package.

Connection Performance. All three methods have good records of reliability, provided that design and process rules are followed and the right protection is used. Protection materials and processes are covered later in this chapter.

1.5.5 Second Level: To PWB

There are two basic categories of connections for second-level, or board-level, assembly of the area array package: mechanical connections and those joined with conductive material. The most common

FIGURE 1.7 TAB (tape automated bonding).

mechanical method uses pins, and the package is called a *pin-grid array* (PGA). Other mechanical attachments have been proposed, and some have even reached prototype stage, but only the PGA has become important to date. For example, a small amount of work has been done on boards that provide plug-ins for BGAs and flip chips with hard bumps such as copper.

Connections designed for joining are either spherically shaped or columnar. The joining material can be the solder obtained by melting the interface or externally applied metal. Conductive adhesives also can be used to mate metal bumps or balls as well as columns. Other chapters will cover many aspects of BGAs.

1.6 PACKAGING MATERIALS

One can view the package as a miniature container that houses one or more semiconductor devices and provides at least the minimum requirements that were listed in Table 1.1. The container can be made of anything that will satisfy the requirements. The protective enclosure can be fabricated before assembly or after. Metal, ceramic, and plastics (polymers) are the three common categories and are now discussed in more detail.

1.6.1 Metal

Metal has nearly all the desirable properties for a packaging material. Metal is a barrier to all gases, including water vapor, one of the most common corrosivity-promoting agents. Metal is also a good heat-transfer material and can be processed into any desirable shape by common methods supported by an immense infrastructure. However, metal is also a good electrical conductor that potentially can short out connections to the semiconductor devices. The last problem is solved by using insulators, such as ceramic or PWBs. The metal package is designed to be evacuated and sealed so gases that could cause problems are reduced to an acceptable minimum. This type of package is referred to as a *hermetic* or *sealed hermetic package.* Special molecular scavengers, called *getters,* can remove contaminants throughout the lifetime of the package. A typical design is more or less that of a small box that is open at the top and finally sealed with a lid. The electronics is sealed within, and electrical leads exit by means of insulators that maintain gas-barrier properties. Alternately, an insulative base made of ceramic can be used to solve the electrical isolation issue. Figure 1.8 shows metal hermetic packages.

1.6.2 Ceramic

Ceramic material has long been used because of excellent barrier properties, good thermal conductivity, and good electrical isolation. In many ways, ceramic is the ideal packaging material. However, ceramics are more expensive, and their processing is not as versatile as that of plastics, which can be molded, fluid dispensed, and even stenciled. Ceramics also are processed at high temperatures that can exceed 1000°C. Yet ceramics enjoy high-volume use especially for larger chips despite the costs. Many CPUs use ceramic chip carriers in pin, column, and ball grid formats. Some of the newer packages use a flip chip attached to a ceramic surface in an FCIP. The low CTE, extremely good flatness, low moisture absorption, and high thermal conductivity make the ceramic carriers very suitable for the fast, high I/O, and hot running computer chips. Other versions have a cavity where the chip is

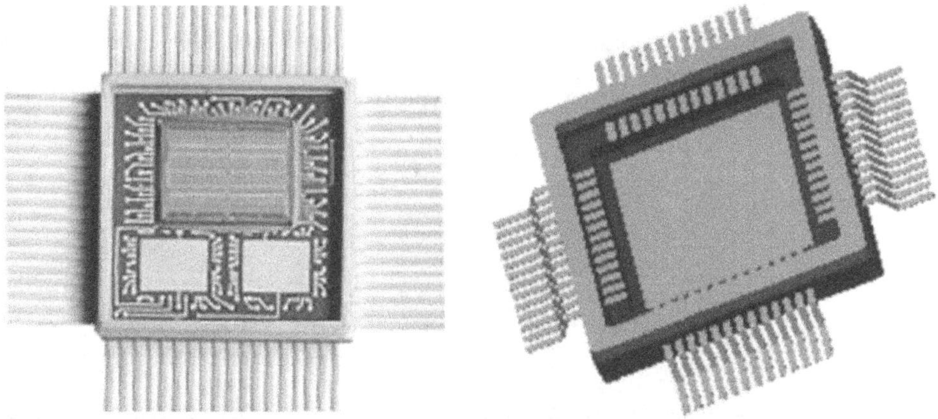

FIGURE 1.8 Metal hermetic packages.

placed and then wire bonded to conductor pads on the ceramic. This type of package can be sealed with polymer encapsulant. Figure 1.9 shows ceramic packages with an attached chip that is ready for sealing with a lid or filling with encapsulant. One must keep in mind that plastic is a good, but not a perfect barrier, so it does not provide a hermetic environment.

1.6.3 Plastic

Most packages are made of plastic because of the significant cost and manufacturing benefits, even though polymers do not provide a hermetic barrier. Polymers are extremely versatile, and there is essentially an endless number of packages that can be made with highly engineered and tailored properties. A key attribute of polymers is that they can be handled easily and shaped efficiently. Thermoplastics can be melted and formed into the desired shape and hardened again by cooling. However, most electronic polymers are thermosets that are hardened to permanent, nonmelting shapes. The thermosets can begin as liquid or low-melting solid precursors that are shaped easily and precisely by any one of numerous industrial processes. Plastic area array packages use several different plastic encapsulation processes that are fast and efficient, hence their popularity. The common starting point is an area array chip carrier that is encapsulated with a thermoset polymer.

Rigid Carriers. Rigid carriers are the most common types of plastic package platform. The carrier is basically a miniature PWB that provides the chip mounting platforms and the wiring interconnect structure. FR4 has been used, but higher-performance organic laminates such as bismaleimide-triazine (BT) have become common. BT and other advanced laminates have more thermomechanical stability and heat resistance. The chip carrier is heated four or more times during its life: die attach, chip interconnect, solder ball attachment, and assembly to board. The carriers typically are thin two- or four-layer circuits that are produced in multicircuit (or carrier) panels for manufacturing efficiency. The panels often are cut into strips that make a convenient form factor for chip-attach, ball-attach, and encapsulation machines. Figure 1.10 shows a rigid organic chip carrier just prior to encapsulation.

Flexible Carriers. Flexible circuitry has been used as a chip carrier since the 1960s, but flex-based packages are increasing in popularity today. Several commercial designs are based on thin copper/polyimide circuits. They may be divided into two categories based on the conductor geometry and the resulting package size compared with the chip. Flex with a fan-out conductor configuration produces a BGA configuration in which the package is larger than the 1.2 times value arbitrarily used to classify CSPs. A fan-in routing, in which the leads track inward under the chip, can generate less than 1.2 times chip area to achieve CSP designation (Fig. 1.11).

FIGURE 1.9 Ceramic area array packages.

FIGURE 1.10 Rigid BGA carrier (before and after encapsulation).

All three types of first-level chip connections are used with flex, DCA, WB, and TAB. The TAB-like connection is the most unusual and does not have an equivalent in rigid systems. IBM was the first to use TAB inner-lead bonding while removing the outer-lead connection. The leads from the chip were routed into an area array pattern and connected to metal bumps by means of vias through the thin polyimide film. This innovation kept the best of TAB and eliminated the troublesome outer leads. What's more, the package was transformed into SMT. IBM named the product *tape ball grid array* (TBGA). The μBGA from Tessera also uses a TAB-like cantilevered beam array for the chip connection. Gold leads are bonded to ordinary chip pads as in TAB ILB, but the outer leads are gone. The conductors fan in underneath the chip and terminate as bumps under the chip to produce a very small area array package. Since the conductors fan in, the package is much smaller than the equivalent TAB style. The flex-CSP can be surface mounted instead of using a specialized TAB bonder. Both the TBGA and micro-CSP are successful.

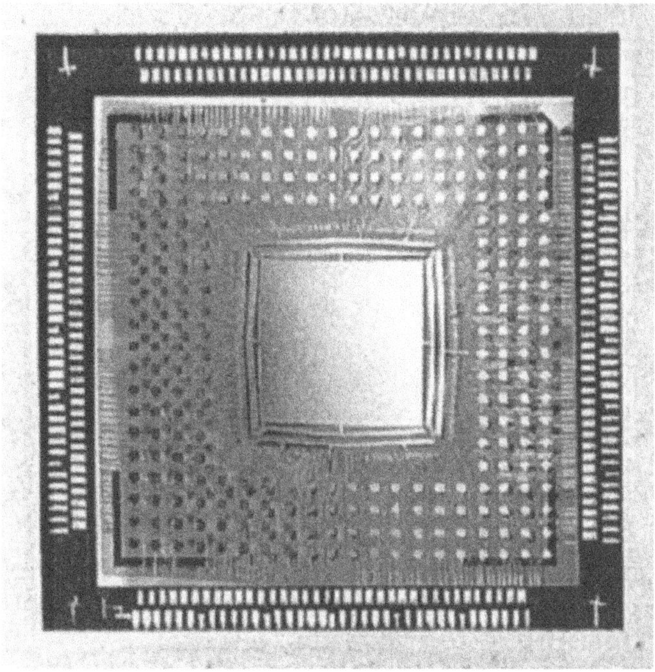

FIGURE 1.11 Flex BGA.

1.7 PACKAGE STYLES

1.7.1 Lead Frame

As suggested earlier, metal lead frames can only be used for perimeter-type packages and will only be mentioned in this book for reference purposes.

1.7.2 Chip Carriers/Platforms

Chip carriers are necessary for area array packages, but they also can be designed as perimeter-only packages. Once again, perimeter types will only be included within this handbook for reference purposes. The chip carrier serves the mechanical function of providing a platform for the chip. The chip is strongly bonded to the carrier with die-attach adhesive or other means. The chip must then be connected electrically to the carrier, which routes the lines to the correct positions for bonding to the PWB.

Perimeter. Carrier-type packages, even with metal ball connections, can be perimeter packages. The larger carrier-type designs are almost exclusively area array packages. However, small CSP designs are often perimeter style. There are two reasons. The simplified connection schemes may not allow area configurations. Some simple CSPs bring the connections to the sides of the package so that no through-the-carrier conductors are needed. The second reason is that low-count packages, such as those for single memory chips, do not always need to use area. This also makes inspection easier because the connections can be examined visually.

Area Array. Area array packaging can be as simple as a double row of connector balls around the perimeter of a chip carrier. Many of the early BGA packages used a full area deployment with continuous rows of balls or columns. The platform area where the chip was bonded adhesively was constrained from normal movement by the low-CTE silicon chip. The organic board behaved like it had a much lower CTE, more like that of the chip. Solder balls on the BGA substrate therefore were placed under greater strain when the BGA was assembled to a PWB. The organic PWB and chip carrier would be expected to expand at similar rates, except for the area under the chip, where natural movement is restricted. This localized differential rate of expansion caused sufficient under-chip stress to produce selected solder ball failure. Subsequent BGA designs relieved the under-chip area, thus avoiding full arrays.

Chip Location. The chip can be placed onto the topside of the carrier in a *die-up configuration* with electrical signals routed to the underside. The other design is *die-down,* sometimes called *cavity-down,* where the die or chip resides in a recessed cavity on the same side as the bumps. The cavity-down design precludes bumps in the center, but this is not an issue, as explained earlier under "Area Array."

Multiple Chips. Multiple chips on single carriers are becoming increasingly common. The *multichip package* (MCP) has achieved what the multichip module (MCM) failed to do. In fact, the MCP is the MCM. Many of the current CPUs have several capacitors on their carrier. Memory packages have been designed with four flip chips. The use of multiple chips is a natural evolution of the advanced area array package, and we can expect continued development here.

1.8 PROTECTION

1.8.1 Hermetic

The hermetic package is a gas-tight container that does not require an encapsulating material. The hermetic package is not a common class for area array packaging and may be viewed as a specialty type for very high reliability or unusual systems.

1.8.2 Molding Compounds

Molding compounds have represented the preferred means of package protection for several decades. The most common materials are epoxy molding compounds (EMCs). EMCs consist of solid epoxy resins, solid hardeners, fillers, and additives. Generally, they are provided in a cylinder shape referred to as a *puck* or *hockey puck.* However, they come in a myriad of shapes designed for different molding machines and run sizes. A transfer-molding machine heats the solid slug of EMC to melting and then injects it into a mold loaded with packages. Heating causes the prepolymer mix to polymerize into a hard, durable, nonmelting mass that permanently takes on the shape of the mold. Area array packages can be molded in singulated form, in strips, or in sheets. A mold can hold hundreds of packages to produce many thousands of parts per hour. More recently, strips and sheets of carrier substrate have been molded for high throughput and mold standardization. The entire carrier sheet containing the attached chips can be flood or area molded and later sawed into separate packages. Not only does this increase output, but standard molds also can be used that preclude having to build complex molds for each different package.

1.8.3 Liquid Encapsulants

Liquid encapsulants typically are made with epoxies and hardeners having a composition similar to EMCs. The polymer precursors are liquids instead of solids like the EMCs, but they can be hardened into polymers with similar properties. Liquid encapsulants typically have less filler to allow flow and do not quite measure up to the EMC properties. Liquid encapsulants are applied with needle dispensing equipment that is slower than molding machines but more versatile. Needle dispensers are programmed to apply the required pattern and therefore do not need tooling. This means that the liquids are ideal for both short runs and quick time-to-market applications. Die-up designs require that a dam first be placed around the chip to retain liquid encapsulant. The dam can be made by needle dispensing thick encapsulant in a pattern. The low-viscosity fill-type encapsulant is then dispensed over the chip. Both the dam and encapsulant are hardened in the same bake process. A die- or cavity-down package is inverted, and the bottom die cavity is filled with a needle dispenser. Recently, high-output dispensers have been offered with up to four heads.

1.8.4 Underfill

Underfill fits within a subcategory of encapsulants but has an additional role to play beyond barrier protection. The underfill acts as a mechanical coupling adhesive that anchors the chip to the carrier and prevents differential movement that could weaken and eventually destroy the solder or conductive adhesive joints of the flip chip. Underfills have become more complex, and at least three types are available with a fourth, wafer-level, film under development. Application methods include needle dispensing and even transfer molding. These are covered in Chap. 6, "Flip Chip Technology."

1.8.5 Chip Passivation Only

Early flip chips had glass or other inorganic passivation over the active side of the die. Bumps were designed to fill in and close any gap so that the flip chip was effectively sealed off from the environment. Passivation is still used today. A passivated chip may not need underfill if there is a reasonably good thermomechanical match between the chip and the substrate, as is the case with ceramic carriers. However, the common practice is to still use underfill for added cycle life and better environmental protection, especially for larger chips.

1.9 PACKAGE ASSEMBLY

Area array packages are surface mountable, although pin-grid array types could be considered through-hole in some cases. However, the more common method of board connection for PGAs is by means of a socket on the PWB. The PGA that plugs into a surface package should not be considered a feed-through type. The PGA, although one of the earlier area array packages, continues to remain popular because its pluggable feature allows easy field servicing as well as convenient upgrading, especially for computers.

1.9.1 Issues with Area Array

The biggest issue with BGAs and other nonpluggable area array packages has been inspection after assembly. The assembly industry had become so accustomed to visual inspection that a package with hidden joints was rejected by many manufacturers. While the issue has subsided to some degree, many people still view the lack of optical inspection as negative. Simplified real-time x-ray equipment has come into use that allows careful inspection, but one can argue that the high cost compared with a magnifier makes this only a partial answer.

1.9.2 Soldering

Solder joint formation is the most common method of assembly for area array packages. Most BGAs have balls or spheres made of eutectic solder so that solder paste need not be applied. Paste can be used, but a more common practice is to only apply flux.

1.9.3 Conductive Adhesives

Conductive adhesives could be used hypothetically for the large BGA packages, and work was done in this area. The concept was to eliminate the solder balls and apply patterned anisotropic conductive adhesive (PACA) film to the underside of the package. The assembly would then be bonded with heat and probably pressure. A major problem was package warp that was considerably worse several years ago. While solder balls can accommodate significant noncoplanarity, adhesive film is much less tolerant. Furthermore, a symmetric BGA creates strong alignment forces that can correct a misoriented package, and this is not the case with adhesives. However, smaller packages, especially flip chips, did not present a coplanarity problem. Both isotropic (ICAs) and anisotropic conductive adhesives (ACAs) are used commercially for flip chip assembly. When used with a nickel or gold bump, the adhesive provides a system that is ideal for memory chips because there are no radioactive emissions. Memory chips and some of the high-density CPUs do not tolerate α-particle emissions that emanate from natural lead in standard solders that contain trace amounts of radioactive nuclides.

1.10 PACKAGE RELIABILITY

Area array packages have equal or better reliability than the equivalent perimeter-style packages. Earlier problems with solder ball fatigue under the die have been corrected primarily by leaving the under-chip area open. The partial array design also relieves density and makes routing easier for the PWB manufacturer There are still some assembly concerns for plastic BGAs because moisture can enter the package through the organic carrier. Absorbed moisture can cause explosive evolution of water vapor during solder reflow heating that fractures the molding or causes delamination between encapsulant and carrier. Either storing BGAs in a dry box or predrying before assembly can eliminate this phenomenon, known as *popcorning*. While these are effective solutions, future materials and designs may eliminate these added procedures. However, the higher processing temperatures expected for lead-free solders could worsen the problem. Ceramic BGAs and CGAs do not have this problem. Also, the higher standoff of the columns used in CGAs nearly eliminates any fatigue in thermocycling conditions.

Flip chips on ceramic may have the highest record of reliability of any system. IBM has assembled millions of flip chip modules over the past three decades with essentially no package failures. However, second-generation flip chips on organic substrate encounter higher stress forces, and failure can occur sooner. The stress-relieving (diverting) solution is underfill. Underfill allows the high-modulus, low-expansion chip to restrain the carrier's typically high expansion. However, during the evolution of underfill and the related processes, problems were encountered. Some underfills lacked the high adhesion required and delaminated from the chip surface, allowing bumps to fracture during thermocycling. Voids in the underfill also were found to cause premature failure depending on their number and location. Newer predispensed underfills that have inherently higher CTEs also may result in lower reliability. However, the majority of underfill-related failures occurred in the laboratory and during development. In general, second-generation flip chips can pass 1000 thermocycles without failure, and the industry considers this acceptable. At the same time, flip chips on ceramic have been tested at high as 60,000 cycles without failure. Flip chips and underfill are covered extensively in Chap. 6, "Flip Chip Technology."

1.11 FUTURE EXPECTATIONS

Area array packaging is expected to continue to gain market share as ICs move to higher lead counts and product designs require further miniaturization. While the target moves slowly upward, better performance and economics are found with area array packaging with lead counts of over 400 for standard fan-out BGA packaging. However, area array chip-scale packages are being used for very low counts for memory chips where small size and high speed are becoming increasingly important. MCPs also will be a growth area.

Future BGAs will continue to shift to DCA for the first-level connection as lead counts grow and the flip chip infrastructure expands. This type of package, FCIP (flip chip in package), is already the preferred style for computer chips because wire bonding and TAB become impractical at the high (>1000) lead counts. Furthermore, CPUs in flip chip format can be designed with redundant power and ground bumps so that power does not need to be routed within the chip with lower-current-capacity submicron conductor traces. Additional bumps can be added for heat piping. The bump can be an excellent path to conduct heat away from the IC into the carrier and finally to the PWB. Additionally, heat can be removed from the back of the chip by applying a heat sink. Faster, more powerful CPUs have an increasing heat burden, and FCIP is one of the best designs for dealing with heat. This also suggests that FCIP for computer chips will continue to leave the back of the flip chip exposed for heat removal.

We will continue to see much progress in the CSP area as more cost-effective products are introduced. While the popular CSPs, such as the Tessera μBGA, are very reliable, many manufacturers seek lower-cost solutions using existing equipment. AMKOR and other packaging foundries, along with large vertically integrated electronics companies, continue to evolve lower-cost CSPs. We can expect a sorting out of the several dozens of styles in the future. Simpler, lower-cost CSPs will become available that perhaps better fit the established processes and equipment already available.

CHAPTER 2
TRENDS/DRIVERS IN THE ELECTRONICS MANUFACTURING INDUSTRY

R. Lasky and Y. Morvan
Cookson Electronics

2.0 INTRODUCTION

As the electronics industry expands and matures, it is essential for those who wish to prosper in the future to understand the fundamental trends that are shaping this most vital industry. It is important to start the discussion of these trends by recognizing that products drive technology, not vice versa. Hence the general trends in products will drive technology and manufacturing needs. The National Electronics Manufacturing Initiative (NEMI) determined that the desire for smaller, lighter, faster, and cheaper products (hereafter referred to as "the fundamental four" or TFF) is the product driving force that will shape electronics manufacturing. Two of the hallmark products that presage these trends are cellular phones and laptop computers. It is little wonder, then, that NEMI chose an advanced personal data assistant (PDA) with a modem as a "product emulator" for its roadmapping and strategic planning exercises.

The electronics assembly capability to effectively support these changes is only now emerging. The current state of the assembly and printed wiring board (PWB) industry can be fairly compared with that of automobile assembly before Henry Ford. As in Ford's day, the continued growth and maturation of this industry will result in tremendous benefits for those who comprehend and take advantage of these opportunities. For those who stay with the old way of doing things, failure will be their lot.

The changes that we see occurring in response to the fundamental four product attributes are the following:

1. *Parallel and data-driven assembly processes.* Up to today, engineers often integrate an assembly line with disparate equipment, often using equipment cost as the main purchasing criterion. The advent of costing software such as SPACE has shown that equipment represents only 1 to 2 percent of the cost of the product produced. SPACE also has shown that increasing throughput with the right equipment can have a dramatic impact on profits. A 4-second reduction in cycle time often can increase profits by more than 25 percent. As this throughput issue is addressed, it will come to be recognized that assembly equipment that is more parallel, data-driven, and flexible is a fundamental part of the solution. Equipment cost will be a secondary consideration; cycle time will be the first concern.

2. *Software and line integration.*
   a. *Software replaces steel/modular, small size.* In the late 1970s, the fighter plane became a computer peripheral. A technology analyst noted this point because at that time, more than 50 percent of the cost of the plane was in computers and software. Assembly equipment is experiencing a similar transformation. The content of assembly machines will shift to software and vision systems replacing massive gantries or turrets and powerful motors.
   b. *Complete software line integration arrives.* Soon, individual assembly equipment must be capable of complete software line integration. This compatibility will be required for the total data-driven processes described earlier. Without this capability, the line will not be capable of the fastest throughputs and changeovers and hence will not be competitive, no matter how low the equipment cost.
3. *Slow secondary processes become niche (TAB, DCA?).* Direct chip attach (DCA) is not a new process; it was invented by IBM in the 1960s and has been used ever since. In 1995, about 450 million DCA dies were assembled, representing only 1 percent of all packages. Currently, DCA may never grow as strongly as many forecast, the reason being that slow, secondary processes are required by the assembler to essentially manufacture the first-level package for the DCA on the PWB. Without material and process breakthroughs to improve the speed of the DCA process, it will never become common. These same issues will drive the package types that will prevail among QFP, BGA, CSP, DCA, and TAB.
4. *Through-hole assembly maintains strong presence through 2005.* Through-hole assembly still accounts for nearly 35 percent of the volume in electronics. The main reason for this staying power is that so many applications do not require the high performance of surface-mount technology (SMT), and through-hole assembly can provide a low-cost solution in these cases. It is also favored in emerging economies because it is an easier technology to implement. Through-hole assembly will continue to generate considerable revenue well beyond 2000, but we must be aware that some of this business will change. An example of such change is through-hole components being assembled in single-center reflow operations. These changes should be viewed as an opportunity for new products and applications.
5. *Nonstandard assembly (optical, odd form) arrives.* Nonstandard and "odd form" assembly is an emerging field. It has been highlighted by NEMI as one of the critical enablers for future high-volume products such as PDAs. Nonstandard assembly in all manufacturing processes represents at least a $100 billion dollar per year opportunity for future nonstandard assembly machines. This opportunity will not continue to be neglected as vision, robotics, and machine intelligence become lower in cost and proficient in handling assembly tasks currently only performed by hand. Optical assembly, although smaller in dollar value at about $10 billion, may be addressed first because it may require less skilled robots. There also exist natural centering mechanisms, discovered in the DCA process, that may facilitate this new field.
6. *Throughput, yield, and cost guarantees. Total solutions.* As pitch becomes finer and input-output (I/O) counts increase, the difficulty in maintaining high yield increases exponentially. This demand and the requirement for maintained and increased throughput as pitch narrows will increase the value of process knowledge. Knowledge will be sold individually or tied into materials and equipment. Manufacturers eventually will want process guarantees from their suppliers.
7. *Knowledge.* It will become more and more clear that knowledge in electronics has a value as processes and productivity become more dependent on knowledge.
8. *PCB/flex line widths from 6 to 3 mils as standard.* The ultrafine-pitch (UFP) technologies that will emerge with chip-scale package (CSP), DCA, etc., will require substrates with line widths and spaces to 3 and even 2 mils. The standard line widths of 5 to 6 mils today will approach and even surpass 3 mils by 2000.
9. *Lead-free issues: Pb/Sn still dominates.* Environmental issues are important, but the difficulties in finding good substitutes for Pb/Sn and the increased realization that electronics only contributes a small percentage of environmental lead will ease efforts for lead-free solders. Hence electroconductive adhesives (ECAs) and lead-free solders will only enjoy niche (albeit important) markets. Were lead to be banned as waste, recycling would challenge lead-free processes as a solution.

10. *Passives continue to be needed, imbedded emerge.* Passive components will not be "integrated away." They will continue to grow at more than 20 percent a year. It is most cost-effective for LSI integrated chip (IC) manufacturers to design a few basic microprocessors or application specific integrated circuits (ASICs) and "tune" the circuit with passives. I/O count, clock frequency, lower-power circuits, and analog applications all drive the need for more passives. The trends in all these "drivers" are to increase the number of passives required. Currently, some applications, such as PCMCIA cards, do not have enough room for all the passives. Buried passives as a solution are being investigated by a DARPA-sponsored consortium. From a PWB, flex, and ceramic materials perspective, this topic offers tremendous challenges and opportunities.

We have reviewed 12 trends that are driving the future of our industry. We believe that these trends will make the future the most interesting, exciting, profitable, and potentially risky period in the history of civilization. We have purposely avoided referring to an eventual maturing of electronics because we believe that it will reinvent and redefine itself throughout the future.

2.1 PARALLEL AND DATA-DRIVEN ASSEMBLY PROCESSES

As the electronics industry matures, the fraction of the value of semiconductors in electronic equipment continues to grow. In the early 1990s, approximately 10 percent of the value was in the semiconductors, whereas today the figure has risen to over 20 percent. In portable devices such as PDAs and camcorders, the semiconductor value has skyrocketed to 50 percent. There has never been an industry so successful as the semiconductor industry in adding value to its products. In 1970, 20 MB of RAM cost $20 million, about as much as a commercial jet. Today, 20 MB of RAM costs less than a coach ticket from Boston to Atlanta.

It is difficult, then, to look to the semiconductor industry to lower the cost of the final product. This situation has increased the pressure on the packaging and assembly industry to control cost. For the first time, the cost per I/O in many packages has dropped below $0.01 per lead. The assembler also shares this cost pressure. Hence the most critical parameter for low-cost assembly is throughput. With quality as a given, throughput is what makes money.[1] It is becoming understood in the industry that the two critical parameters to ensure optimal throughput are that processes be parallel and data-driven.

2.1.1 Importance of Throughput

Although the intent of this section is to discuss parallel and data-driven processes, profit is so strongly tied to throughput that it should be discussed first as background for a thorough understanding of these two critical parameters. To understand the effect of throughput on profit, we will analyze the assembly of a typical unit—a low-price CD-ROM player. To start, consider the important elements of cost[2] in Table 2.1.

The cost parameters can be calculated with a recently developed cost-estimating spreadsheet. Table 2.2 lists some of the cost output from this scenario.

When costs are analyzed at the level of the assembly line (i.e., costs only calculated with labor for the assembly-line workers), the nonmaterial assembly cost per I/O, often called the *conversion cost,*[3] is considerably less than the industry standard at the factory level. This situation exists because there are considerable numbers of workers who support the product but do not work on the assembly line. These include designers, administration, purchasing, hand assembly, shipping, etc. In the simulation in Table 2.1, adding the 26 "overhead" or indirect workers provides an industry competitive conversion cost of $0.0144 per I/O. Strictly speaking, many of these 26 people are in manufacturing and would not be considered overhead in most analyses, but they do not work on the assembly line for this analysis.

To see the effect of throughput on profit, assume that the selling price is $110, and calculate the gross profit as a function of cycle time. For the baseline case (25 seconds = 0.416 minutes), the gross profit per year is approximately $4.5 million. The cost spreadsheet is able to plot the gross profit

TABLE 2.1 Assembly Cost Elements

Item	Cost
Assembly equipment	$1.7 M
Floor space + utilities	$40,000/yr
Components	$75/unit
Stencil, squeegee	$600, $200, life = $1,000,000
Solder paste, consumables	$0.7/unit
PWB	$15/unit
7 workers	$28/ h (includes benefits)
Selling price	$110
Shifts	12 10-h shifts per week
Cycle time	25 seconds
Downtime, setup, maintenance	8%, 10 h/wk–12.5 h/wk
Workers supported	26 @ $42/h each
I/O	950
Yield	97% first pass, reworkable

TABLE 2.2 Cost Analysis

Element of cost	Cost per board ($)	Percent of total cost per board
Labor (direct and indirect)	12.44	12.0
Machine depreciation/repair	0.75	0.72
Floor space and utilities	0.06	0.06
PWB and components	90	87.0
Rework	0.35	0.30
Consumables (paste, etc.)	0.18	0.17
TOTAL	**103.78**	**100**

versus cycle time, illustrating the stunning effect of cycle time on gross profit (Fig. 2.1). Reducing the cycle time from 25 to 20 seconds (0.33 minute) increases the gross profit to almost $8 million per year! The reasons for the increase are twofold: First, more product is produced, and second, unit costs are reduced. These two factors result in an exponential increase in gross profit as cycle time is reduced.

This analysis also suggests the path to financial success in subcontract assembly. We have been in many discussions during which someone has said, "Aren't you glad your business is not subcontract assembly? The gross margins are so small." It is true that the gross margins are only 3 to 10 percent, but this point misses the essence of financial success in this business. Although the margins are low, the throughput can be so high as to produce an almost unequaled return on assets. Returning to the example, assets were about $2 million (assuming perhaps some asset valuation for the building, e.g., $300,000 per $4.5 million) If net profit is 50 percent of gross, net profit is then $2.25 million. Hence return on assets (ROA) is 112.5 percent! This ROA is clearly world class. From the previous discussion it is now clear that from an ROA perspective, ROA grows strongly with throughput.

2.1.2 Parallel and Data-Driven Processes

Parallel processes are processes that accomplish several process steps at the same time. Examples might be stencil printing (many pads printed simultaneously), "walking beam" pick-and-place machines (placing several components at the same time), and arguably, reflow ovens (reflowing all components at once). The throughput benefit of parallel processes is obvious and universal. *Universal* means that a wide variety of process lines benefit from parallel processes. Hence, for a dedicated line, the process steps being parallel usually will enhance productivity in the same product day in and day out. Even if your line produces a wide variety of products with frequent changeovers, parallel processes are still attractive.

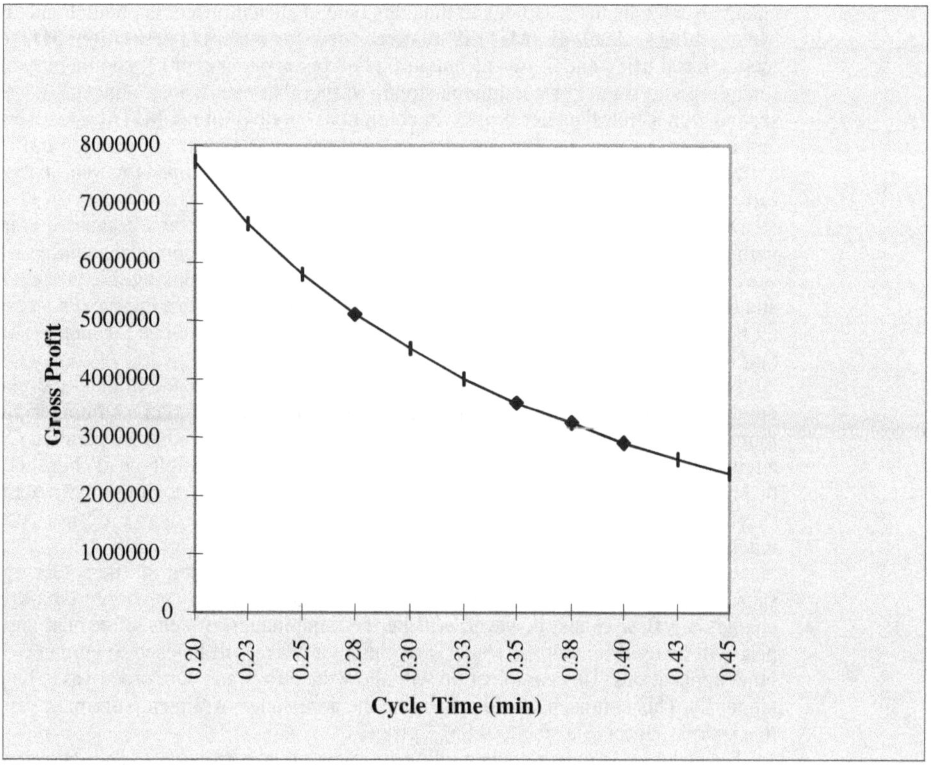

FIGURE 2.1 Gross profit versus cycle time.

The opposite of parallel processes is *serial processes.* An example of a serial process is a single-beam, gantry pick-and-place machine. From a throughput perspective, the disadvantage of serial processes is self-evident: It is difficult for something to be fast when it is done one step at a time.

Data-driven processes only require the change of a software program to assemble a different product. Examples of data-driven processes are changing the software program on an adhesive dispenser or pick-and-place machine. The advantage to throughput is less time lost in product changeovers.

Hard-tooled processes are the opposite of data-driven processes. An example of a hard-tooled process is stencil printing. The benefit of data-driven processes depends strongly on the number of product changeovers required. If there are few product changeovers, data-driven processes are not as critical to throughput. If frequent changeovers are the rule, data-driven processes can greatly improve throughput.

2.1.3 Developments in Parallel and Data-Driven Processes

Recently, there have been many efforts to develop and improve assembly processes so that they are both parallel and data-driven (P&DD). Unfortunately, few process steps have both these characteristics. Simultaneously, much R&D currently is being pursued in this direction. The following is a review of a few current developments that do fall into the P&DD realm in solder-paste printing, dispensing, pick-and-place, and reflow ovens. The most dramatic changes in parallel assembly are anticipated in pick-and-place machines

Stencil printing of solder paste is usually parallel and hard-tooled (one stencil per product design), but some current machines[4] will handle up to six separate stencils. This feature allows product changeovers to be more rapid, and thus this type of stencil printer is parallel and "near" data-driven. Metal jetting technology (MJT) is relatively new for materials deposition. MJT deposits drops of molten metal using one of two techniques: (1) drop on demand or (2) continuous stream.[5,6] It is serial (one drop at a time) but continuous stream jetting that can deposit almost 1 g/s (10,000 drops per second with 8-mil-diameter drops). In continuous-mode printing, MJT is a fast serial (near-parallel) and data-driven process. Drop-on-demand technology is slower but simpler and more mature. It is anticipated that the first manufacturing-capable MJT will be shipped this year in the drop-on-demand format, with continuous mode to follow.

Dispensing of adhesives typically is a serial data-driven process because it is usually performed with one dispensing head. However, product changes can be accomplished with the change of a software file. Dual-head dispensers are currently on the market,[7] making this process now near parallel and data-driven. The dual heads allow the dispensing of multiple adhesive dot sizes, enhancing overall throughput. This trend is expected to continue with multiple-bead dispensers in the not too distant future.

The most dramatic changes in parallel assembly are anticipated in pick-and-place machines and are data-driven because product changeovers only require changing a computer program. However, more effort will be spent on producing pick-and-place machines that are parallel. Currently, there are a few such offerings, most notably the "walking beam" and multiple-head chip shooters.[8] In the drive to produce parallel pick-and-place machines, it will become obvious that machines with 100-kg overhead gantries to place parts that weigh less than a grain of rice do not constitute optimized machine design.

Reflow ovens are clearly a parallel process, in that they melt all the solder on the board at the same time. There are also significant efforts to make reflow ovens more data-driven. The biggest change in reflow ovens, however, will be the combination of convective heat and microwave. The heat will be used to melt the solder, and the microwave will be used to cure epoxies in flip chip or other applications. This combination will allow the reflow and cure processes to be performed simultaneously. This approach would accelerate the acceptance of flip chip dramatically and would make this process closer to a standard SMT process.

The industry is only now beginning to realize that both parallel and data-driven processes are needed to achieve the productivity goals of the future. Eventually, we should expect to see process lines with all processes being parallel or data-driven.

2.2 SOFTWARE AND LINE INTEGRATION

In the 1970s, the jet fighter became a computer peripheral. A technology historian made this interesting observation. His reasoning was that at that point in time, greater than 50 percent of the value of the fighter was in computers and software. This trend certainly has continued in the aircraft industry. Many planes today cannot be flown without computers.

We expect that this trend will be more and more evident in electronics assembly equipment also. In the near future, tremendous increases in productivity and efficiency will occur by combining the increased use of computers, vision systems, and software on the individual assembly machines and the integration and control of the entire assembly line with "system" computers.

2.2.1 Software and Machine Vision Replace Steel

Software and machine vision improvements will affect most of the equipment used in PWB assembly. However, pick-and-place equipment likely will be affected the most, so we will use this machine as the basis on which to elaborate.

The pick-and-place process, as currently practiced in a typical fine-pitch placement machine, is not optimized from either a machine or process perspective. To understand these perspectives, we will look at this process in some detail.

Step 1. As the pick-and-place process begins, the machine first loads the PWB.

Step 2. The machine then finds several global fiducials on the PWB so that it can have the necessary references to properly place the components at the correct location. The machine knows the locations for the components because the PWB layout had been entered previously into the controlling computer's memory.

Step 3. The machine then picks the component from a feeder *and aligns it with respect to the placement head* in preparation for placement.

Step 4. As the machine places the components, it measures the location to place them by referencing from the location of the fiducials. This step requires a precise (and expensive) scale. *In the time domain, this step can be performed in parallel to step 5, but it is delineated separately because of its importance in relating to the complexity of machine design.*

Step 5. The component is then moved into position. The PWB has dimensional stability to about 0.001 to 1.0 in, so for finer-pitched components [say, 0.4-mm quad pack flats (QFPs)], an additional fiducial near the placement location (called a *local fiducial*) may be required to achieve proper placement. The machine has to reference to this fiducial before placing the QFP, thus taking additional time.

Step 6. The machine places the component.

These steps are shown pictorially in Fig. 2.2.

This approach has several fundamental disadvantages. Referencing to the PWB global fiducials is the most obvious. The lack of PWB dimensional stability is a critical shortcoming, making such referencing nonoptimal. This lack of dimensional stability can be such that it is not possible to place a component with ultrafine-pitch leads accurately enough. In such cases, a fiducial near the component placement area is required. The placement machine must then find this local fiducial and take the time to perform an additional positional referencing operation. Hence, as ultrafine-pitch packages become more and more common, more and more local fiducials will be required to place them, slowing the placement process down more and more. The advent of area array packages, with their coarser lead spacings (such as BGA and CSP), will only temporarily relieve this concern because in time, even these packages will evolve to finer pitches. The additional time to reference to these local fiducials will be unacceptable in an era of increased productivity demands.

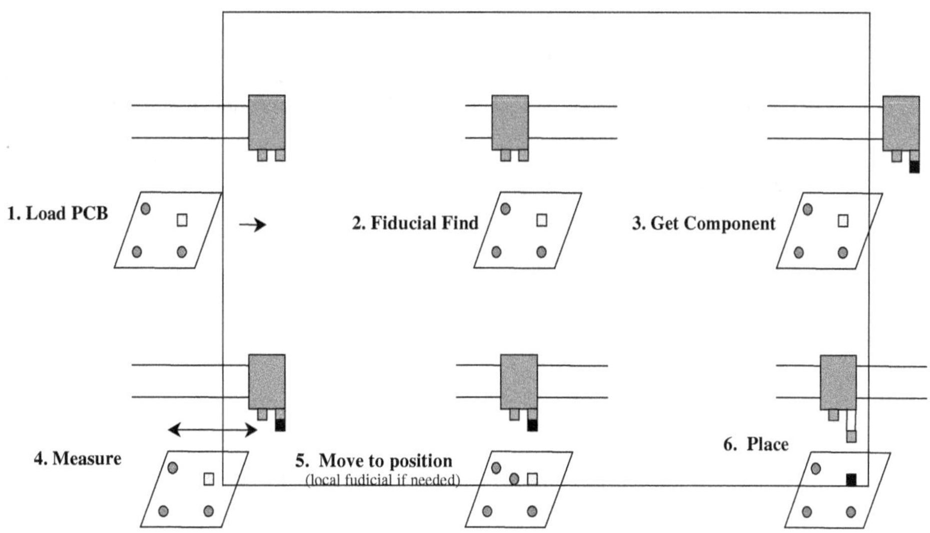

FIGURE 2.2 The steps in pick and place.

In the placement process, an expensive scale is used to measure placement positions relative to global fiducials. To obtain accurate measurements, the entire machine frame must be very stiff. This requirement may result in an overhead gantry weighing upward of 100 kg (over 200 lb) in this type of pick-and-place machine. This massive overhead gantry then requires powerful (20-hp or so) and expensive motors to drive it. These designs can result in machines costing over $300,000 to place 5000 or so components per hour. It is interesting to contemplate that when placing a 0402 passive capacitor, the gantry weighs 100 million times more than the passive (100 kg versus 1 mg). In no other technology are such ratios tolerated to deliver such a small payload.[9]

The National Center for Manufacturing Sciences (NCMS) chaired an effort called Light, Flexible, Mechanical Assembly (LFMA) in the early 1990s. The purpose of this work was to develop assembly equipment that could perform the complex assembly tasks currently only performed by humans (screwing a lid on a jar and assembling the covers on an electronic product such as a personal computer are examples). This effort spawned a small company, CAMotion,[10] that is working on this complex problem. Optimized pick-and-place machines are a simpler subset of this challenge, and this topic also is being addressed by CAMotion. This new machine-optimized approach will now be discussed. A schematic is shown in Fig. 2.3.

As with current machines, step 1 involves loading the PWB into the machine. In step 2, the machine picks the component from a feeder *and aligns it with respect to the placement head* in preparation for placement. In step 3, the machine places the component without any fiducial referencing. The machine knows approximately where the placement site is from mechanical information entered into the machine with the computer-aided design (CAD) data. It aligns precisely to the placement site by imaging the pads, *referencing to them,* and then placing the part.

This approach not only reduces the steps to three but also eliminates the need for precise scales and a stiff, exacting frame. The resulting machine would have the equivalent of a 10-lb gantry and a 1-hp motor.

Several technology enhancements are needed before the look-while-place approach is realized. The first is the availability of low-cost, light, and compact vision systems. The vision system will need to be in the 1 × 3 × 2 in range and weigh less than 1 lb. The second need is the development of machine learning algorithms that can "learn" what the typical placement sites look like, say, a QFP or BGA site. The last technology need is design improvements in the control of flexible structures. Reducing the mass and stiffness of the placement machine will enable a tremendous cost and speed performance advantage. These advantages have a price, however. The lighter, less stiff frame likely will be subject to greater vibration problems during placement. These vibrations can cause a lack of placement accuracy. The science of the control of flexible structures is maturing to a point where machines can be designed so that such vibrations are readily not excited. In addition, learning algorithms can be designed that not only recognize the placement position but also learn how to move quickly to the site while exciting minimal vibrations. We believe that these types of technologies will have a profound and perhaps disruptive effect on the assembly industry in the not too distant future.

These three technology challenges are currently being pursued, and it appears that the current approach, as described in the next subsection, has promise. However, even if this approach proves unsuccessful, another will be developed to perform look-while-place action—the prize is too valuable to ignore.

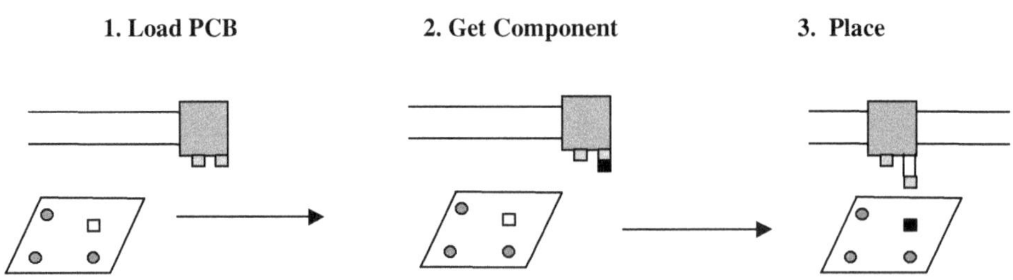

FIGURE 2.3 Look-while-placing approach.

2.2.2 Line Integration

Integration of the assembly line with software and computers is becoming common and will become ubiquitous in the near future. Productivity will be the main "driver" of this trend. This topic is broad and goes well beyond our scope here, but we will mention some of the current line-integration products that operate under the most common architecture, GEM-SECS (Generic Equipment Module–Semiconductor Equipment Communications System). This architecture was developed for semiconductor fabrication equipment and was adopted for assembly equipment. GEM-compliant architecture allows one to configure different assembly machines with different operating systems into one computer database system. Hence your dispenser, printer, and oven may run Windows 95 and your placement machine may run DOS, but they can still communicate if they are GEM-compliant.

Recent products such as GEMHost and GEMComm[11] are examples that work within this mantle. They improve integration and allow better scheduling, line balancing, statistical process control (SPC) data collection, and cost control. At this time, most of these products only allow line communication for data collection but do not enable complete line control from one computer. In the future, such total line control and even control at the factory level will become common.

2.3 SLOW SECONDARY PROCESSES BECOME NICHE

As we discussed the importance of parallel and data-driven processes in PWB assembly earlier, we emphasized the importance of productivity to profit. Thus it should not be a surprise that any process that slows down the cycle time of the assembly line erodes profits exponentially. Tape automated bonding (TAB) was a classic example of this phenomenon. Through the late 1980s and early 1990s, TAB was the "package of the future." TAB's low profile and very fine lead spacings (down to 6-mil centers) seemed to give it numerous advantages. The manufacturing of TAB packaging was performed with many pieces of equipment that were "off the shelf" because they also were used for the manufacture of 35-mm photographic film. Hence TAB packaging would appear to have many advantages: It is inexpensive to manufacture and has the finest lead spaces and lowest profile. Thus TAB was poised to take the world by storm, but it didn't. Understanding why will be very instructive in predicting which future component technologies will prevail.

An understanding of TAB's downfall comes through an understanding of the assembly process. Let us assume that a TAB package is a candidate to replace a BGA package in the example that we analyzed previously. Please refer to Table 2.1 again for a list of pertinent information.

Let us assume that the TAB substitution saves $2 on the bill for materials. What a terrific start! We again use the cost-estimating software SPACE[12] to perform the cost analysis. At first glance, it appears that we have reduced the cost by almost 2 percent. The bad news is that TAB packaging requires two new TAB machines that cost $350,000 each, and the cycle time must be slowed down to 30 seconds. These changes result in the cost and profit figures shown in Table 2.3. The results from the previous analysis (BGA) are listed for comparison.

By using the TAB process, the unit cost actually goes up by more than 1 percent, but the most stunning change is that the gross profit goes down by more than 30 percent from $4.14 million to $2.66 million. The reason is that the number of boards produced dropped from 661,914 to 551,595. Again, the profit is hit by the "double whammy" of reducing productivity—fewer units at a slightly higher unit cost. SPACE indicates that if we add yet another TAB machine and it can bring the cycle time back to 25 seconds, the profits are then restored with a slight increase. Unfortunately, the addition of three machines and the additional space and workers needed may present other challenges unrelated to cost.

This analysis indicates a strong fundamental principle. *Lasky's rule: Any process change that slows the process down usually will result in lower profits. Corollary: Slow secondary processes almost always should be avoided.*

Hence flip chip or DCA also should be a productivity concern because it has several added process steps. Typically, flux must be dispensed onto the substrate at the chip site. The die then must

TABLE 2.3 Comparison of TAB and BGA costs.

Element of cost	BGA build ($)	TAB build ($)
Labor (direct and indirect)	12.44	15.39
Machine depreciation/repair	0.75	1.23
Floor space, electric, etc.	0.06	0.078
Materials (components, PWB)	90	88
Rework	0.35	0.32
Consumables (solder paste, etc.)	0.18	0.18
Total cost per board	103.75	105.18
Number of boards produced	661,914	551,595
Gross profit	$4.14 million	$2.66 million
Delta gross profit		−$1.48 million

be placed with a pick-and-place machine that has a special head. After reflow, the die then must have underfill dispensed under it, and the underfill must be cured in a curing oven. Hence additional equipment in the form of a modified pick-and-place machine, a dispenser, and a cure oven is required. Evidence of this process slowdown exists in that it has been stated recently that the DCA process requires 38.5 percent more time in cellphone assembly. Thus it is likely that DCA will only become a mainstream technology when these processes become more like the standard SMT process. Work in this regard is currently underway.[13]

2.4 THROUGHPUT, YIELD, AND COST GUARANTEES: TOTAL SOLUTIONS

We have discussed the need for continuously improving throughput previously, so we will not emphasize it again. An additional challenge to throughput, however, will be the trend that good yields will be harder and harder to achieve. The main difficulty in maintaining good yields is that yield loss grows exponentially as pitch is reduced. To understand this situation, let us assume that yield loss per lead varies as in Eq. (2.1).

$$YL = a \exp\left(-\frac{b}{p}\right) \tag{2.1}$$

where YL is yield loss per lead, a is a constant, p is the pitch in mils, and b is also a constant. This type of relationship might be obtained from process data, where the process capability is constant and the only variable is component pitch. The constants a and b would be determined from the process data. Let's say that a given pager assembly process capability results in 1 ppm yield loss per lead in a process when the lead spacings are 65 mils and 20 ppm when the spacings are 20 mils. Usually, Eq. (2.1) and a little algebra reveal that $a = 0.265$ ppm and $b = 86.5$ mils. Applying these constants to Eq. (2.1) results in a process capability as shown in Fig. 2.4. To be fair, this analysis is highly idealized. The lead spacings in any product are not of only one value, and process capability is not as easy to model as we have shown. However, the general trends that are shown here (i.e., that yield loss per lead increases strongly as pitch decreases) are inarguable.

Compounding the preceding concern, as pitch decreases, the number of leads per area in a given package increases. It can be shown that this relationship will increase roughly as

$$P = \frac{c}{p^2} \tag{2.2}$$

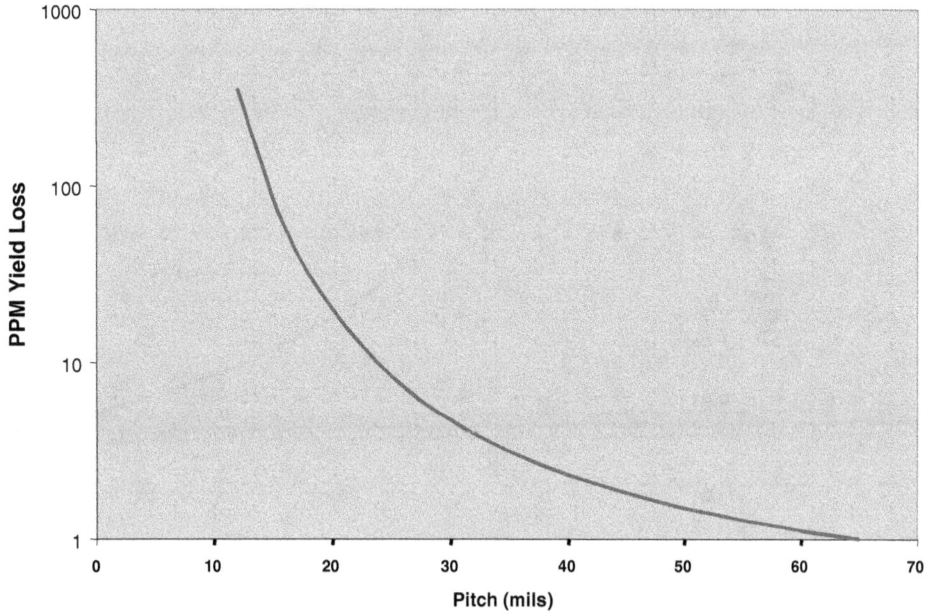

FIGURE 2.4 Yield loss (ppm) versus load spacing.

where c is a constant, and P is the number of PWB pads (i.e., leads) in the package. Thus, for an old pager technology where p might equal 25 mils and P is 150, c then equals 93,750 (i.e., 150×25^2). If the technology is reduced to 16 mils, Eq. (2.2) suggests that P then equals 366, assuming the same size package. Again, this calculation is a "ballpark figure" used to show a trend, since the packages have a mix of pitches.

Typically, a manufacturer will be concerned with end-of-the-line yield loss, i.e., how much product does the line lose during the entire process. End-of-the-line yield loss is the loss per lead times the number of leads:

$$YLEOL = a \exp\left(\frac{-b}{p}\right) \times \frac{c}{p^2} \tag{2.3}$$

In our fictitious pager example we have determined all the constants and can now plot Eq. (2.3). This plot is shown in Fig. 2.5. This figure indicates that in our pager example the end-of-line yield loss might be only 0.1 percent when the lead spacings are 25 mils, but they skyrocket to 2 percent as the lead spacings are reduced to 16 mils. Figure 2.5 suggests that if the lead spacings are reduced much below 12 mils, the yield loss will be catastrophic.

Although this analysis is fictitious, the thrust of the argument is not. As lead spacings are reduced, yield loss per lead increases, and there are more opportunities (i.e., more leads) to be at risk. Hence the combined result is that as pitch decreases, end-of-line yield loss increases exponentially if process capability is not improved.

The preceding analysis suggests strongly that process improvements will be necessary just to hold yield loss stable as pitch decreases. Area array technologies such as BGA may relax finer pitch requirements temporarily in some applications. However, with the continuing trend to produce smaller, lighter, faster, and cheaper products, even BGA eventually will have finer and finer pitch requirements. As one improves one's process capability, one must not forget Lasky's rule. Do *not* slow the process down.

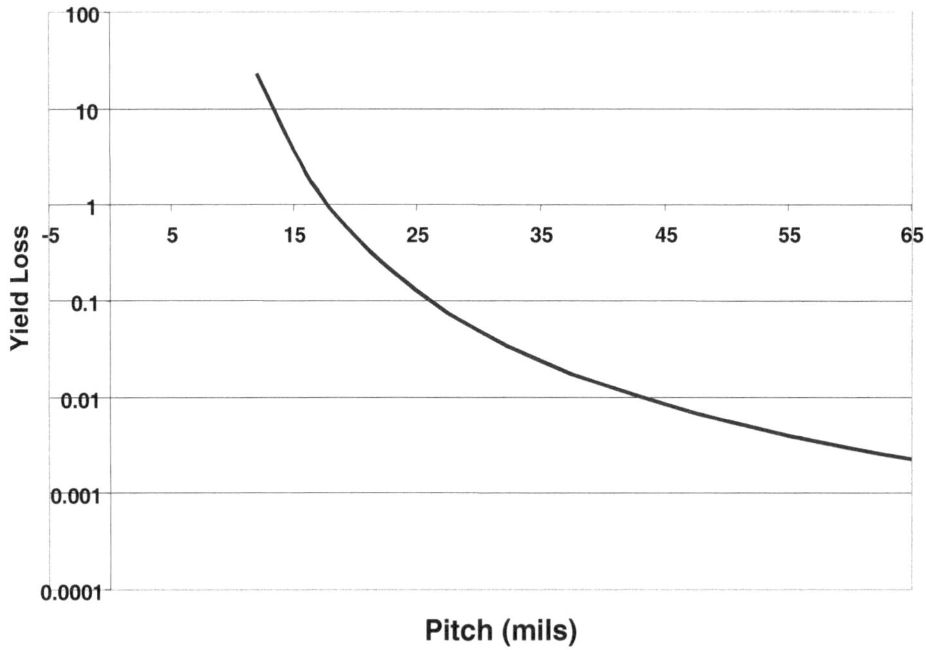

FIGURE 2.5 Product yield loss.

The twin and somewhat mutually exclusive challenges of greater productivity and improved yields can only be met by a thorough understanding of all individual (i.e., printing, placement, etc.) and systems (i.e., the effect of solder-paste parameters on the entire process, etc.) aspects that affect productivity and yield. It will be more and more difficult for assemblers to meet the twin challenges of improved productivity and yield unless they work very closely with their material and assembly equipment suppliers. This need will encourage the offering of integrated solutions (i.e., tailor-made combinations of materials and process equipment and process knowledge to fit the assembler's needs). Such offerings are now emerging.[14] Ultimately, it may be that customers demand process guarantees. Such guarantees undoubtedly will require an unprecedented closeness between assemblers and their vendors.

In conclusion, both productivity and high yields will continue to be required to compete in the electronics assembly marketplace of the future. Companies that survive and prosper will master assembly system techniques that enable greater productivity and higher yields. It will not be acceptable to trade one or the other off. Such needs will drive the offering of integrated solutions in the future and will elevate the value of process knowledge in its own right.

2.5 THROUGH-HOLE ASSEMBLY REMAINS AND ODD-FORM ASSEMBLY EMERGES

2.5.1 Through-Hole Assembly

It is interesting to reflect on the changes that have occurred in electronic assembly in the past generation. It started with armies of people on an assembly line placing single resistors, capacitors, inductors, and transistors. Today's machines can place greater than 75,000 components per hour with accuracy approaching 50 μm (2 mils). Early efforts to automate assembly were pioneered by such companies as Universal Instruments and IBM. The resulting through-hole assembly equipment concepts developed in these early years are still used today, accompanied by many innovations. The rapid acceptance of surface-mount technology (SMT) can give the false impression that through-hole assembly is "going the way of the buffalo." Figure 2.6 shows the number of though-hole and SMT components projected to be assembled. It should be noted that as recently as the early 1990s, more through-hole than SMT components were assembled. Even in the year 2000, about 25 percent of all components were through-hole.

The reasons that through-hole technology is only fading slowly are straightforward: cost and simplicity. Cost is the most obvious strong point for through-hole technology. We can all get excited about SMT and flip chip and BUM (built-up multilayer) PWBs, and they are clearly needed for laptop personal computers (PCs) and "handycams," but we must remember that a large portion of electronics is simple. An example might be a garage door opener. This small portable unit is inexpensive and relatively low tech. It can be manufactured with through-hole technology in a low-labor-rate region of the world such as China or Mexico. There is little reason to change to SMT for products such as this, especially since it would be more expensive.

Electronics has been the industry of choice for third-world countries with strategic economic growth plans. The reason is simple: Electronics historically has grown four times as fast as economies in general [i.e., 8 percent composite annual growth rate (CAGR) versus 2 percent for economies as a whole]. When a country does choose to enter electronics, the simplicity of through-hole technology makes it a natural choice. We know of examples where a small through-hole assembly line was set up in a house in a third-world country. One only needs axial and radial inserters and a wave solder machine and one is ready for business. In one such "facility," the kids played beneath the solder pot! In contrast, SMT typically requires some experienced technicians and operators and is a considerably more demanding process.

2.5.2 Through-Hole Equipment

The keen-minded process engineer who embraces today's modern arsenal of manufacturing equipment is usually geared for the assembly of advanced technology products. To many professionals of this audience, the following may not at first resemble legitimate reasons to use through-hole technology. However, we all must remember that at least in part, ingredients for productivity

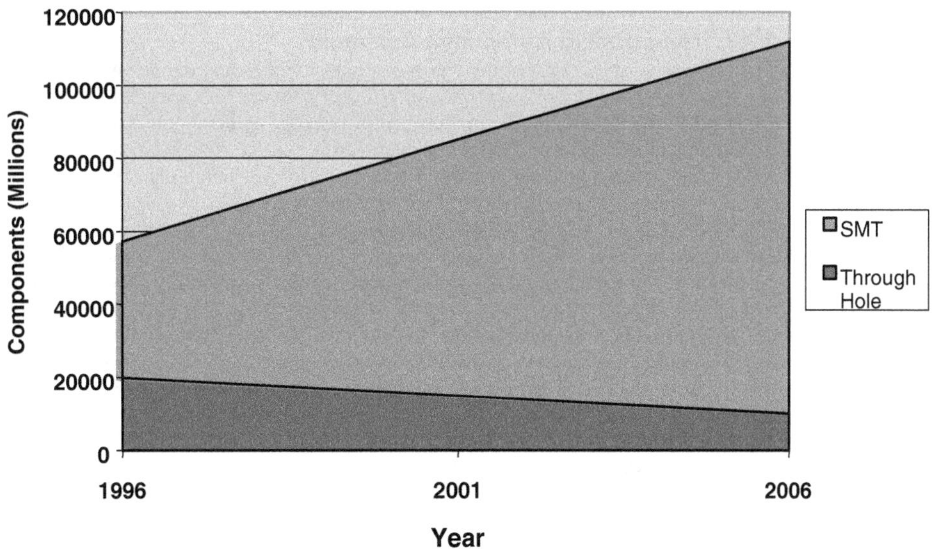

FIGURE 2.6 Through-hole and SMT component consumption.

include the manufacturing environment, the quantity of units needed to be produced, and end-product requirements.

As compared with SMT, through-hole component technology assembly is more forgiving to harsh manufacturing environments, more adaptive to batch-style manufacturing, and more conducive to manual assembly; the PWBs and components generally are cheaper; and the technology is less process-intensive. On the other hand, one reason this technology faces demise is the lack of component availability as SMT takes an ever-increasing hold. This is ironic because a decade ago, SMT was slow to get started for the same reason (lack of available components in its own domain). Yet SMT assemblies still must yield to the older through-hole technology. Most PWBs today employ both technologies. Many devices such as connectors, transformers, inductors, large electrolytic capacitors, and high-power components are through-hole in nature.

This is not intended to imply that through-hole technology assembly techniques and equipment have sat still. The complexion of automated through-hole assembly has changed dramatically. For instance, at one time it was necessary to presequence components on tape prior to the insertion operation. More modern equipment allows component sequencing, electronic verification, and insertion all on the same machine. The equipment also has become more flexible and can accommodate a much wider breadth of component specifications such as lead spacing, lead thickness, and body configuration. In most cases, the newer equipment is able to tolerate component leads that are more deformed. Historically, these machines also were stand-alone in nature and required an operator to manually mount and remove PWBs from workboard holders as a function of the process. This did not fit the trend of more modern approaches in that only batch manufacturing was possible. Even though some assembly houses still use these techniques, most modern assembly equipment facilitates "in line" manufacturing with pass-through board-handling systems. This approach reduces operator dependency and allows the PWB assembly to pass between process cells with conveyor systems instead of humans. In most instances, the dedicated workboard holder has been eliminated.

A major supplier in Japan offers a high-speed VCD axial-lead component-insertion machine with an insertion speed of 18,000 pieces per hour. This is twice that of the company's previous model. In addition, with two separate component feeder carriages, components can be exchanged and replenished during operation. A major supplier of equipment in the United States offers a radial insertion line rated at over 11,000 components per hour. This equipment handles an even wider variety of components than its predecessor, which before were inserted at a rate of about 4000 parts per hour.

2.5.3 Nonstandard Automated Assembly

The impressive state of assembly equipment and robotics and the recent chess victory of Deep Blue can give the wrong impression of the abilities of robotics and artificial intelligence (AI). Assembly equipment with its associated robotics, vision systems, and computers can accurately place thousand of components per hour on PWBs and even verify passive values. However, the most complex robot with AI and vision cannot perform the simplest task for a 3-month-old baby—to recognize her mother's face from across the room. Many of us were happy for IBM when Deep Blue became world chess champion, but we will be really impressed when computers can understand poetry and "read" nonverbal communications between people. This introduction sets the stage for the challenge of performing odd-form and final assembly with machines. Assembly robots are outstanding at rapidly assembling standard components that are simply precisely placed on a PWB. However, the assembly of multiple odd-shaped components is a great challenge due to difficulties in grasping and correctly orienting the component. Even a greater challenge for machines is final assembly and placing the finished product in a box. It is difficult for machines to master the spatial recognition and mechanical movement that a person easily performs in the final assembly of a VCR or other typical electronic product.

Presently, SMT is at somewhat of a mature state, *mature* in the sense if excluding the normal invention of related advanced processes, components, and process materials of the technology. One can say that SMT has become more "customary versus revolutionary." However, we cannot ignore the influence that other component technologies has on the SMT process, such as leaded through-hole components. The impact that through-hole and odd-form devices have on the overall SMT process can be quite significant. Many companies, consultants, and trade organizations have addressed the solder attachment of these components using conventional SMT process materials and equipment. The primary process prescribed is the use of solder paste. The paste supplies the solder volume and flux required to form the solder joint. The through-hole device is then concurrently reflow soldered with the SMT components. This process has been referred to under various names and acronyms such as *pin-in-paste, reflow of through-hole, solder paste on through-hole technology* (SPOTT), *alterative assembly reflow technology* (AART), and a host of others. Unfortunately, various issues are usually encountered: "Can the device withstand the solder paste thermal reflow profile?" "How do we get enough paste on the PWB?" "How and at what stage of the process do we mount the device on the PWB?" In many instances, a workable process can be obtained. However, because of the inherent uniqueness of each assembly, the manufacturing process should be addressed on an individual basis.

Economic opportunities for nonstandard assembly. The economic opportunities in nonstandard assembly are overwhelming and somewhat difficult to quantify. However, we will make an attempt to do so. It has been estimated by Yann Morvan of Cookson Electronics that there are approximately 25,000 SMT assembly lines in the world. Combining this number with the information in Fig. 2.1 would suggest that approximately 11,000 through-hole assembly lines exist. Hence the total number of assembly lines is in the 35,000 range. Assuming a minimum of 10 people (quite low most likely) per line performing nonstandard and final assembly and packaging, these tasks represent a 350,000-person industry. Assuming that one $200,000 machine could replace four people (the machine works faster and for three shifts), there is a $17.5 billion market for these types of assembly machines.

This analysis does not even consider the enormous market for these types of machines in other industries. The opportunities in optical assembly alone add another several billion dollars to the preceding figure. In all industries where assembly is done, the market is clearly over $100 billion for such equipment if it existed. This prize undoubtedly has encouraged such efforts as CAMotion (Atlanta, Georgia). CAMotion is a small startup company with ties to Georgia Tech that is working on the technical issues relating to nonstandard assembly challenges. The company hopes to have prototype machines in a year or so.

2.5.4 Conclusion

Through-hole assembly still will be significant in the foreseeable future. The fact that it is a rugged, low-cost process will ensure its life for low- to medium-technology products. Nonstandard and final assembly and even packaging products in shipping boxes are a field with almost limitless potential that is only now being tapped by automation. We expect this field to continue to see innovations in automation as assembly robots master the challenges of these assembly processes.

2.6 KNOWLEDGE

Predictions *circa* the 1980s:

1. The theoretically fastest speed for a modem is about 14.4 kB/s; no modem will ever be faster over a standard telephone line.
2. The theoretical limits of photolithography will never allow less than 0.3 micron -μm line widths in semiconductors.
3. It will never be possible to develop a low-cost PWB technology with pads over vias, less than 4-mil lines, and more than 50 pads per square centimeter that is fabricated in a sequential process.

The reality *circa* 1998:

1. Standard modems now reach 56.6 kB/s, while asymmetric digital subscriber line (ADSL) is a technology that is capable of 1.5 MB/s over standard phone lines! It is expected to be a consumer reality soon.[15]

2. Scientists and engineers are currently working on photolithography tools to manufacture ICs with 0.13 -μm line widths.
3. See Sec. 2.7.

The breaking of theoretical barriers occurs for very practical reasons. In the first example, the installed telephone infrastructure is worth hundreds of billions of dollars. If a technology can be developed that will use the installed hardware base and achieve high data rates, the financial reward will be tremendous. Hence individuals have a very strong incentive to try to beat the theoretical limits. The second example is similar. The next step beyond photolithography is x-ray lithography. This technology requires a synchrotron electron accelerator to generate the necessary x-ray intensities. The synchrotron is expensive ($100s of millions) and dangerous (exposure to its x-rays for a few seconds would be deadly); hence there is a tremendous incentive to break the theoretical barriers here too.

2.6.1 Knowledge: A Most Valuable Asset

All the breaking of barriers described earlier occurs by the application and development of new knowledge. As assembly and related processes become more and more challenging, the value of process knowledge also increases exponentially. This is especially true as it relates to yields and throughput, as we saw earlier. Yet it is our belief that the commodity in shortest supply in our industry today is practical yet advanced process knowledge. We make this statement after numerous tours and audits of assembly lines throughout the industry. In many cases the process engineers, technicians, and workers are assembling UFP, BGA, CSP, TAB, or other new or challenging packages and struggling with the process. Or perhaps they are implementing a new process such as the pin-in-paste process (also known as *reflow of through-hole,* which essentially strives to eliminate the wave solder and/or manual soldering operations even though through-hole parts are still on the board) and do not know where to begin.

All these trends make process knowledge more valuable than ever. Yet at this time the assembly industry has very tight profit margins, and companies often cannot afford to perform the necessary process R&D and collateral in-house training of their employees. Numerous consortia, alliances, and university centers[16–19] have risen to this challenge. These efforts have been documented in detail elsewhere.[20] All these worthy efforts, to some degree, lack one critical component, however, and that is "hands on" process education. We know of one large consortium that investigated a certain process. When the consortium finished, one of the companies hired the consortium leader to implement that very process in its factory. The irony of this situation seems to have escaped both parties. Why didn't the consortium participant implement the process on its own? Was it because of the inability to grasp the knowledge or concept? Or was it lack of internal resources to implement the new process?

We strongly believe that hands-on process training is the most important type of education and will be most sought after in the future to meet the needs of the industry. Hence, if your company needs you or your team to learn the pin-in-paste process, you should go to a hands-on workshop on the subject taught by some industry leader such as Jim McLenaghan.[21] In addition to Jim's effort, MicroTech has been a champion of hands-on state-of-the-art process workshops.[22] There is even currently a hands-on consortium being developed with regard to electronics assembly productivity.[23]

Probably the most important message in this chapter is the value of hands-on education as the most effective way to stay ahead of the competition. Another valuable way to obtain knowledge is by performing literature searches on technologies before investing your valuable R&D time. ITM will perform literature searches on PWB assembly topics for those who wish this service.[24]

And now to answer reality number 3 as listed earlier, an example of the value of knowledge follows.

2.7 PWB/FLEX LINE WIDTHS FROM 6 TO 3 MILS AS STANDARD, MICROVIAS, AND BUM PWBS PROLIFERATE

Through the 1980s, expensive high-tech products such as mainframe computers usually drove advances in packaging technology. A classic example of this is IBM's thermal conduction module

(TCM) and supporting PWB. This technology, released in 1981, is still used in mainframes today. The combination of high I/O density and heat dissipation was the envy of the industry when it was released, and with its enhancements, it is still impressive today. The economic commitment that IBM made to develop this technology was in the billions of dollars.

Today, a significant change is occurring on the packaging development horizon. The consumer product development response to the mantra of faster, cheaper, lighter, and smaller is the main driver of packaging technology. Products such as cellular phones, pagers, camcorders, Personal Computer Memory Card Industry Association (PCMCIA) cards, and laptop and palmtop PCs are requiring new technologies to satisfy market demands for smaller, lighter, faster, and cheaper models. Technologies such as chip-scale packages (CSPs), ultrafine-pitch quad flat packs (UFPQFPs), tape automated bonding (TAB), ball grid arrays (BGAs), and 0603 (60 × 30 mils), 0402, and even 0201 passive resistors and capacitors are all responses to these consumer product demands for smallness. Substrates (typically PWBs) are needed to support the wiring required by these very I/O-dense packages. The standard 5- to 6-mil line, large (>8 mils)- via PWBS cannot wire these types of packages without an excessive number of board layers and will never have enough space for the needed mounting pads for the components. These high-layer boards also would be prohibitively expensive and add extra weight and package thickness.

In addition, these product requirement changes have occurred very rapidly. Figure 2.7 is a graph of connecting pads per square centimeter versus time. Note that the historical trend in a state-of-the-art (SOA) product has had quite a gentle increasing slope. Both National Electronics Manufacturing Initiative (NEMI) and the Semiconductor Industry Association (SIA) were prescient in their predictions that the actual need for pads would skyrocket when related to this historic trend. The first product to show this high pad count was the Sony handycam in 1997. In response to these needs, fine-line PWB, microvia, and built-up multilayer (BUM) PCB technologies have been developed. These new technologies will be reviewed briefly in the following subsection.

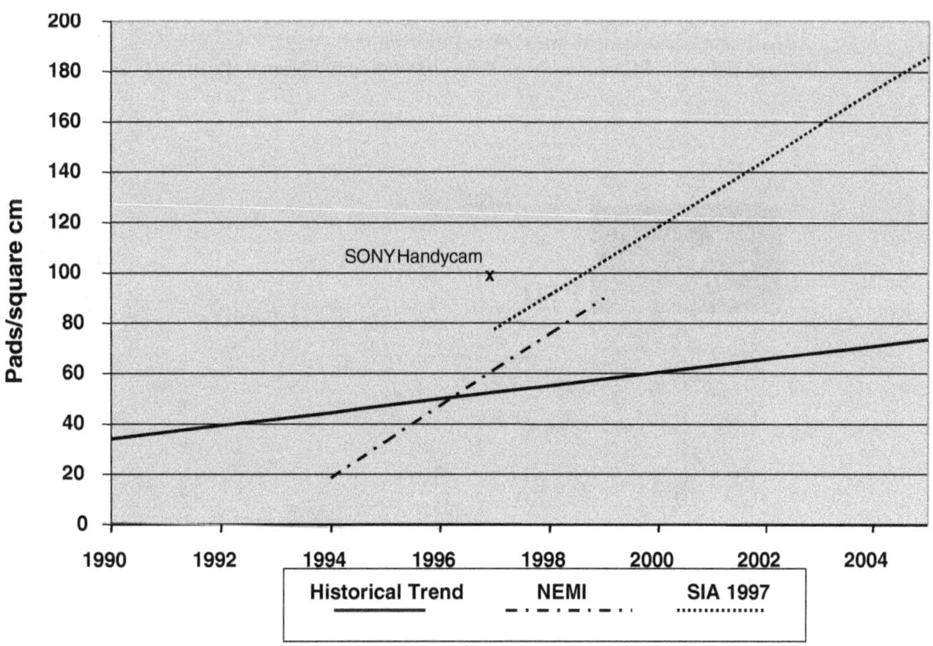

FIGURE 2.7 PWB pad density versus time.

2.7.1 Multilayer versus Built-Up

Multilayer technology has been with us for years. Much of the early work was pioneered by large OEMs such as IBM and Lucent. The driver for the development of this technology was the need for enough wiring and interconnection to support high-performance components and systems. Don Seraphim was one of the pioneers in developing the methodologies for assessing wireability needs for high-performance OEM products.[25] The basic technology is quite straightforward. A schematic of a multilayer PWB is shown in Fig. 2.8. The basic PWB process is as follows:

1. Copper foil is bonded to prepreg.
2. The copper is photolithographically processed to develop the circuit lines needed for that layer.
3. The different layers are laminated and drilled at appropriate levels to form interlayer vias. The lamination is a sequential process (i.e., the layers are laminated one at a time versus one grand lamination at the end).
4. Photolithographic and plating processes are required to create the electrical connection for the vias.
5. The plated through holes are not created until all layers are laminated. The plated through-hole (PTH) process requires drilling through the entire board and plating the hole.
6. The PTH plating operation usually requires another photolithographic process.

This entire process is sequential in that the board is built up one layer at a time. The sequential nature of this process can result in considerable end-of-the-line yield loss. This is true because in sequential processes such as this, net yield is the product of yields at each process step. Hence, if there are 10 steps in a process and the yield at each step is 98 percent, the net yield is 81.7 percent (0.98^{10}).

Conventional multilayer technology has the additional disadvantage that the vias and PTHs take up too much "real estate" that cannot be used to mount components. Although the schematic in Fig. 2.8 is not precisely to scale, it gives the correct sense of the lack of room for attachment pads and channels for wires because so much space is taken up by the PTHs and vias. This situation results in a maximum of about 20 pads/cm^2 on the top or bottom surface of the multilayer board. As noted earlier, currently, consumer products exist that require up to 100 pads/cm^2, and almost all SOA cellular phones, pagers, camcorders, PCMCIA cards, and laptop and palmtop PCs have more than 20 pads/cm^2. Hence multilayer boards, although they could

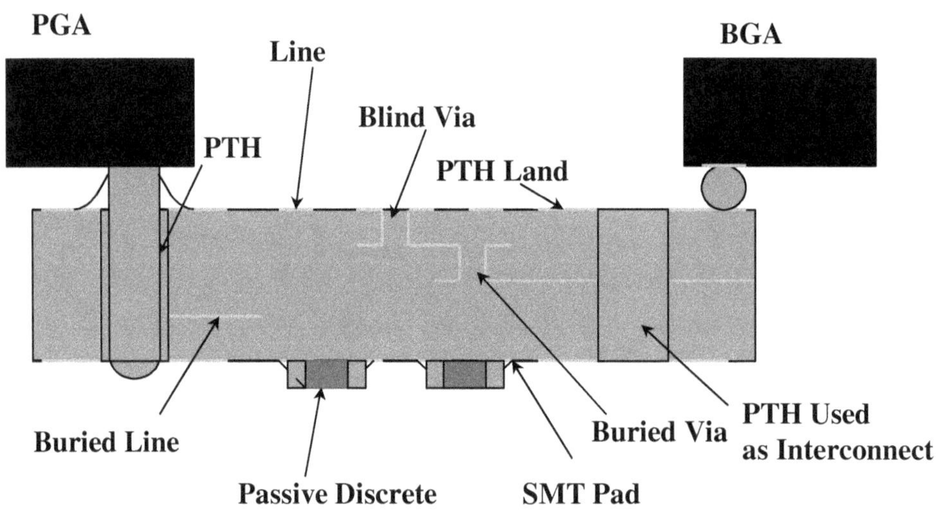

FIGURE 2.8 Standard multilayer interconnection.

have the wireability needed (i.e., length of wire), lack the number of pads to mount components on the surfaces.

In answer to this need, built-up multilayer (BUM) boards were developed. A near-to-scale assembly schematic of a BUM board is given in Fig. 2.9. This technology can achieve greater than 100 pads/cm^2. There are numerous competing processes in building BUMs; a generic process is as follows:

1. An individual prepreg layer is drilled with a laser to form the via holes.
2. Via holes are filled with conductive paste.
3. Copper foil is laminated onto the prepreg.
4. Photo processing forms the circuit lines.
5. The individual layers are tested.
6. The individual layers that form the BUM are laminated together.
7. Photo processing forms the lines and pads on the outer layers.

As noted earlier, the process is not sequential in that the individual layers are not laminated together until they are all circuitized. Since they are tested before lamination, the only yield loss is at the final lamination step. This yield loss typically is quite small. The main functional advantage of BUM over standard multilayer PWBs is that the space that the via takes up can be used by other vias and pads. This single advantage will make BUM boards the technology of the future for substrates. Prismark has contributed several excellent articles on all aspects of BUM technology in its publications for clients.

There are numerous types of BUMs today, and it would be difficult to predict which ones will prevail. In general, though, the BUM technology that achieves lowest cost will be the one to dominate. One thing is for sure — BUM technology increasingly will be the future in substrates. It cannot lose with the combination of a low-cost PWB fabrication process, high fabrication yield, and the greatest pad density and wireability.

There also will be a need for 3-mil line widths at a low cost for some technologies. Currently, line widths below 5 to 6 mils usually exact a cost premium. This need is probably not quite as great as the need for BUM technology, but with flip chip and CSP proliferating, the need for 3-mil lines cannot be too far away.

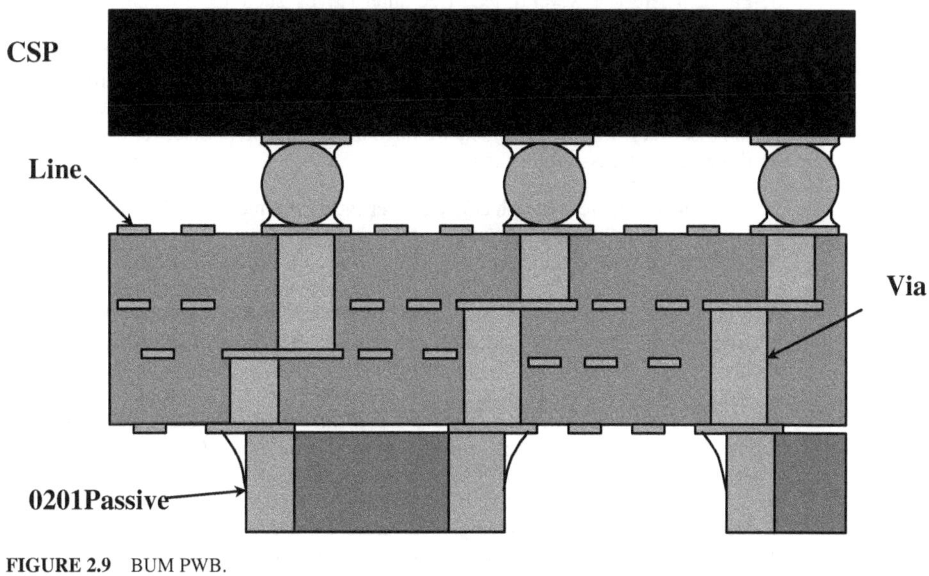

FIGURE 2.9 BUM PWB.

2.8 LEAD-FREE IMPACT

The art of soldering predates written history. Soldering was performed initially with silver, gold, and tin and typically was used as a lower-temperature attaching technique for precious metal castings. By 500 B.C., the Romans had determined that tin and lead produced a very low temperature eutectic solder that could be used to join lead for plumbing and bronze artwork. Hence some of the early cases of lead poisoning were Romans who drank the water from aqueducts. For those who are interested, Alpha Metals has an excellent article on this history on its Web site.[26]

Tin-lead solder has lasted all these years because tin and lead are inexpensive and abundant, melt at a low temperature, "wet" copper very well, and are relatively well understood. The primary negative aspect of tin-lead solder is the toxicity of the lead. There is much concern and confusion today with regard to this topic. Hence we will review some of the issues. Let's start with a true/false test.

True or False

1. Electronics is a major contributor of environmental lead.
2. The level of lead in the blood of U.S. residents has increased since 1945.
3. There are numerous cost-effective alternatives to tin-lead solder.

Answers

1. False. Battery production consumes about 4 million tons of lead, whereas electronics production consumes only 18,500 tons.[27] A significant minority of people in electronics know the qualitative nature of this fact. However, few people know that electronics is not even number two in lead consumption because the manufacture of bullets consumes 200,000 tons of lead. Hence electronics consumes less than 0.5 percent of the world's lead. These numbers do not even include all the lead used in weights to balance tires.
2. False. The level of lead in the blood of U.S. residents has experienced a stunning drop since 1945, going from close to 30 μg/dl in 1945 to less than 5 μg/dl in 1990[28] (Fig. 2.10). The likely reason for this stunning drop is the gradual elimination of lead pipes and lead paint in the environment. The additional steep drop since 1975 also may relate to the use of lead-free gasoline.
3. False. There are several lead-free alternatives such as Sn96.5/Ag3.5 for solder paste and Sn99.3/Cu0.7 for wave soldering. However, the tin-silver solder-paste alloy is twice as expensive, and both of them have melting points that are about 40°C higher than eutectic tin-lead. These and other critical issues in finding lead-free solder alternatives for electronics assembly have been reviewed thoroughly in an excellent article by Bastecki.[29] Unfortunately, in the 2500 years since the discovery of tin-lead solder, we have not been able to discover a metal system that has all its desirable features at a low cost.

Why, then, is there so much concern over the electronics contribution of lead to the environment? For one thing, the numbers in question 1 are slightly misleading. Much of the lead in batteries is recycled, but if only 5 percent of batteries are not, the amount of battery lead is still 10 times that of electronics. In addition, the discarding of old electronics devices is a most visible process. We all discard many electronic items per year. Still, all in all, it is difficult to conclude that the concern for the electronics contribution of lead does not have an emotional content to it. However, as people are inclined to say today, "perception is reality."

Recycling is an alternative to outright banning of lead in electronics. Recycling is already being practiced in some European countries. It typically works by charging an additional fee when the product is purchased; the fee is used to pay for the recycling. One of the issues in recycling is that a by-product of the process can be some toxic gases when the PWB is incinerated.

With all this information as background, we will go out on a limb. We do not think that lead will be banned in electronics in any country. If it were, the difficulty that that country would then have competing in the world electronics industry would be so great that its electronics industry would suffer to the

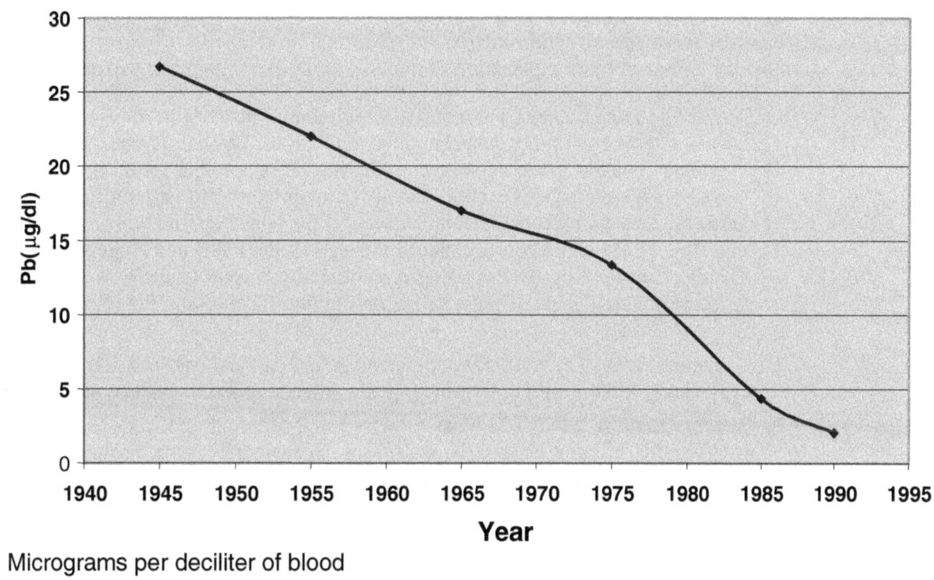

FIGURE 2.10 Average U.S. lead levels in blood.

extent that the legislation would be rescinded. Hence the only way banning could work is if it were done worldwide. We believe that this action is highly unlikely because of the information just stated. We do believe, however, that recycling will become common and even ubiquitous as a response to the just concern for lead in the environment. Recycling will be the answer to the call for banning lead. This does not mean that lead-free solders will not exist and prosper in important niche markets. One possible market is for companies that want to emphasize the environmentally friendly nature of their products. Nortel recently funded the development of a lead-free assembly process for cellular phones for just this purpose.[30] It will be interesting to see how the market receives such products. It also should be pointed out that there are already some significant applications of lead-free conductive adhesives. These adhesives are used in chip-on-glass and other applications where polyester flexible circuits are used. In chip-on-glass applications, typically a die is electrically connected to glass substrates such as in LCD displays. The polyester flex circuit applications are low-cost replacements for polyimide flex circuits. These circuits cannot be processed at reflow temperatures; hence conductive adhesives are used.

2.9 PASSIVES: THE GROWTH CONTINUES

Passives are defined as devices that do not provide power gain. They are most commonly resistors and capacitors but also include inductors and diodes. Since resistors and capacitors are the "lion's share" of passives, the remainder of this discussion will focus on them. Thus, when the term *passives* is used in the remainder of this chapter, it will mean only resistors and capacitors.

The use of passives continues at an astounding rate. A billion hours ago the stone age was the future, a billion seconds ago the iron age was maturing, a billion passives ago was this morning! Currently, more than 800 billion are used per year. This amount is more than 100 for every person on the planet, and if they were place end to end, they would circle the globe more than 20 times! In the future, some people have assumed that passives will be "integrated" away into the IC. The exact opposite is happening—passive growth continues unabated.

2.9.1 Why Is Passive Use Growing?

Passives serve several vital functions in electronic products. In layperson's terms, they are used to tune the circuit. A brief explanation of the most prominent functions performed by passives follows:

1. *Decoupling capacitance.* The constant "switching" that occurs in digital circuits creates noise that can interfere with the power supplied to the numerous circuits on the PWB. If the power supply voltage has too much noise from switching, it is possible for a digital "one" to be read as a "zero," or vice versa. This event is usually quite unlikely, but with millions of operations per second in modern microprocessors, it can occur if the power-supply voltage is not protected. To minimize this effect, capacitors are placed throughout the PWB and charged to the line voltage. The capacitors respond by either adding or subtracting current to the power grid in response to any perturbations to it. This effect is called *decoupling* (decoupling the effects of the switching from the power grid).
2. *Line termination.* One of the requirements of an electrical transmission line is that it be properly terminated. Without proper termination, electrical signals can reflect and cause false signals.
3. *Pull-up resistors.* Many ICs have numerous and different applications. These different applications require different electrical connections. Occasionally, some terminations on the IC will not be used in a specific electrical product. In most cases, the I/O termination in question needs to be set at the power grid voltage with no current flowing. This operation is performed with a large-resistance-value pull-up resistor.
4. *Filters.* Passives are often used to form electrical filters.

It is logical to ask if passives must be used to solve these design challenges. Apparently some of the passive function, especially related to electrical filters, could be integrated into the IC. However, this type of design would limit the IC's flexibility for many uses. Hence there is typically one IC design for many uses, and the circuit is "tuned" for the specific applications. However, it must be understood that even in the best of IC designs, passives that are separate from the IC are needed.

2.9.2 What Are the Trends in Passives?

Current trends in electronics are driving the increased use of passives in all the categories mentioned in the last subsection except for electrical filters. These trends are the need for faster clock speeds, low operating voltages, higher I/O counts on active devices, and increased digital-analog function combinations. Electrical filters can require fewer passives as voltages decrease, but more decoupling capacitors are needed as voltages are reduced. The reduction in filter requirements is usually quite minimal and is overshadowed by the soon to be mentioned growth in other passive uses. Figure 2.11 indicates the stunning increase in PC microprocessor speed over the last 15 years. This same chart shows the typical operating voltages of the microprocessors. These two trends independently require more passives. Together with the increase in active component I/O, they demand exponential passive growth with time. Hence we can expect the growth of passives to continue for the foreseeable future.

Figure 2.12 indicates that the percentage of passives in all SMT components in a PWB has grown from only 25 percent in 1984 to over 90 percent today. This change is arguably the most stunning change in all of electronics. This dramatic change not only requires that more passives are placed, but it inhibits miniaturization, since the passives require more and more area on the PWB. The industry has responded to this change in several ways. First, high-speed chip shooters have been developed. Currently, some chip shooters approach 100,000 placements per hour. A faster chip shooter is a logical first step in the challenge of exponential passive growth. The passives industry also has responded to the situation by producing continuously smaller passives, as shown in Fig. 2.13, and developing integrated passives. These integrated components have numerous capacitors or resistors in one package. This approach also reduces the assembly cost in that only one passive component needs to be placed to satisfy the function of numerous individual passives. Since assembly costs often are quoted as cost per placement, this approach can have a seemingly dramatic effect on cost control. On the negative side, however, these integrated passives are typically of a custom design;

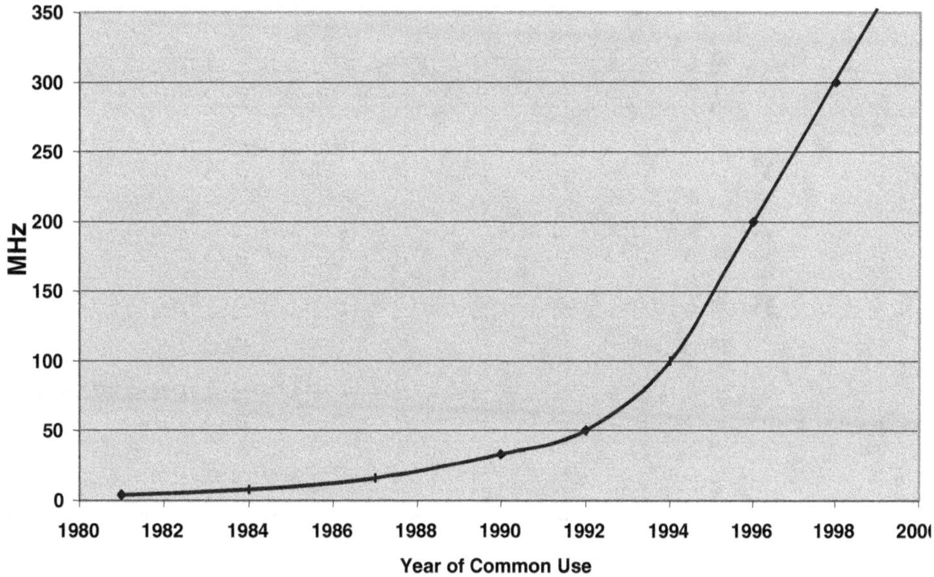

FIGURE 2.11 PC microprocessor speed increases.

hence their unit cost is high. Currently, only about 250 million integrated passives per year are used versus almost 900 billion discrete passives.

In some cases, the number of passives required is so large that there is not enough room on the PWB to place them all. To address this need, considerable work is being performed to develop the technologies required to "bury" the passives in the PWB or flex circuit material. Hadco has been a leader in buried passives in PWBs, while Sheldahl has been a champion of buried passives in polyimide flexible circuits. The buried approach is not particularly flexible from a design perspective and is still quite expensive.

2.9.3 What Is the Future for Passives?

Integrated and buried passives will hold important but niche positions in the entire passive market because of their lack of flexibility vis-á-vis discrete passives. As prices come down, integrated passives will likely grow more than buried passives because their inflexibility is less. We see difficulties in discrete passives becoming common below the 0201 (20 × 10 mils or 0.5 × 0.25 mm) size because their small size creates assembly handling difficulties. Even today, handling of 0402 passives is a challenge.

Thus we see discrete passives continuing to grow, but the 0201 and 01005 sizes being slow to emerge unless placement breakthroughs occur. Integrated passives will grow rapidly from their much smaller base, but they will never seriously threaten discretes. Buried passives will continue to be very much a niche but important solution in the future. They will be used primarily when "real estate" concerns demand their use. We believe that other approaches such as jetting of resistive material are still a long way off.

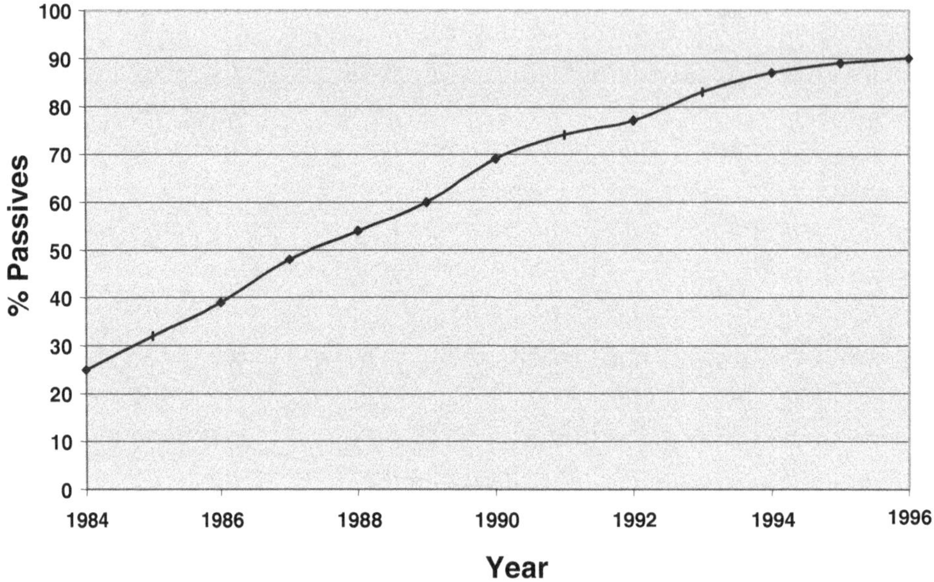

FIGURE 2.12 Passive-SMT ratio in packages.

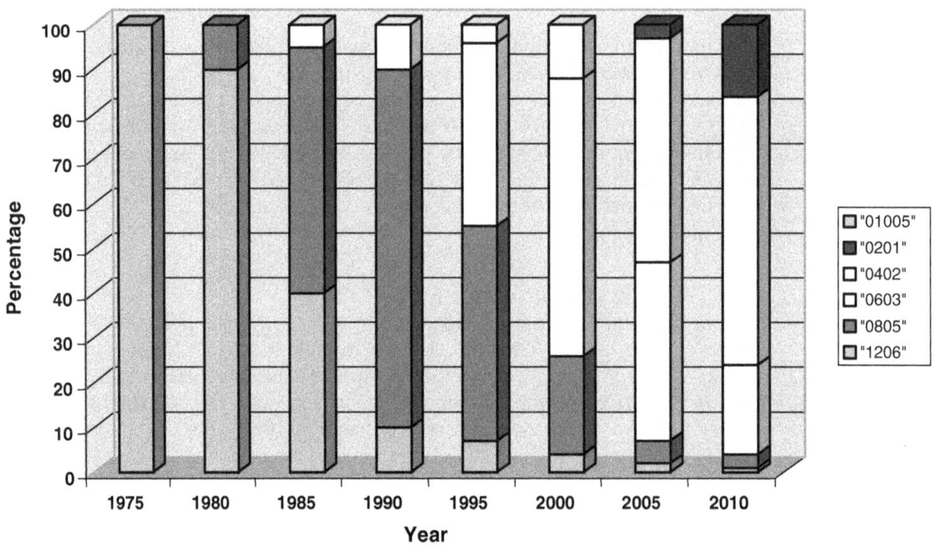

FIGURE 2.13 Percentage of passives by size.

2.10 NOTES AND REFERENCES

1. Bagley, Tom, MPM Corp., private communication, 1996.
2. Lasky, R. C., and Baldwin, D., "Emerging Packaging Technologies," SMTA Conference, Raleigh, NC, November 1996, pp. 1–4.
3. *National Electronics Manufacturing Roadmaps,* NEMI, Herndon, VA, pp. 41–51, 1996.
4. MPM UP3000 Literature, "Solder Jet Printing," MPM Corp., Franklin, MA, 1998.
5. Snyder, M., and Lasky, R., Materials Research Society Meeting, San Francisco, April 1995.
6. Godlin, R., *SMT,* January 1997, pp. 66–68.
7. Camelot Gemini Literature, Camelot Corp., Haverhill, MA, 1999.
8. Phillips FCM Literature, Philips Corp., Eindhoven, The Netherlands, 1999.
9. Another light-hearted view of the situation is that if all the energy in the 20-hp motor were used to move only the passive, it would move it at about 300 times the speed of sound (i.e., it could cross the United States in 8 minutes).
10. CAMotion, Inc., Manufacturing Research Center, Room 436, Atlanta, GA, 30332-0405, Steve Dickerson, President.
11. Speedline MPM, 16 Forge Park, Franklin, MA 02038.
12. SPACE was developed by MPM Corp. A derivative product, CoastCoach, is marketed by ITM, Durham, NH (*www.itm-smt.com*).
13. Professor Dan Baldwin, Georgia Institute of Technology, Atlanta, GA; Steve Corbett, PolyFlex, Cranston, RI.
14. Cookson Electronics, Foxboro, MA (508-541-5800).
15. *PC Magazine,* Feb. 10, 1998, p. 28.
16. George Westby, UFP and BGA/DCA Consortia, Universal Instruments, Binghamton, NY (607-779-5258).
17. Rao Tammula, PRC, Georgia Tech, 813 Ferst Drive, Atlanta, GA 30332-0560 (404-894-9097).
18. NEMI, 2214 Rock Hill Rd., Suite 110, Herndon, VA 20170-4214 (703-834-2082); e-mail: *jmcelroy@nemi.org.*
19. A registry of university alliances can be purchased from HTA, 26 Howe St., Medway, MA 02053.
20. Lasky, Ron, "Consortia, Alliances, Centers and Other Such Phenomena: A Perspective," *Circuits Assembly,* March 1998.
21. Pin-in-Paste Workshop, Jim McLenaghan, Cookson Electronics, Providence, RI 02903 (401-521-1000).
22. Martin Barton, MicroTech, 1416 Glastonbury Drive, Plano, TX 75075 (972-424-8805).
23. Dan Baldwin, Georgia Tech (404-894-4135); e-mail: *daniel.baldwin@me.gatech.edu.*
24. ITM (603-868-1754); e-mail: *theitmteam@aol.com.*
25. Seraphim, Lasky, and Li, *Principle of Electronic Packaging,* McGraw-Hill, New York, 1989, pp. 59–67.
26. *http://www.alpha-polymers.com.*
27. Prismark Partners, 130 Main Street, Cold Spring Harbor, NY 11724 (631-367-9187).
28. *Wall Street Journal,* September 16, 1993.
29. Bastecki, Chris, *SMT,* "Lead Free Solders," May 1997, pp. 52–58.
30. *Los Angeles Times,* December 1, 1997.

CHAPTER 3
AREA ARRAY PACKAGING

Ken Gilleo
Cookson Electronics

3.0 INTRODUCTION

Area array packaging is the oldest and the newest high-density package style. The early IBM solid logic technology (SLT) package developed in 1961 appears to have been the very first area array design. SLT was used for transistors, which were the only commercialized solid-state devices at that time. The array was the simplest possible because only three inputs-outputs (I/Os) were required for the common transistor. Soldering three copper microspheres to the transistor pads created the micropackage. The package was passivated with glass, but once it was assembled by solder reflow, an underfill was applied for environmental protection. The SLT was used in the famous System 360 mainframe computer launched in 1964. Figure 3.1 shows the SLT. Note that this is not only a ball grid array (BGA), it is also a chip-scale package (CSP) and the first surface-mount technology (SMT) product.

Modern area array packages include BGAs, CSPs, and flip chips (FCs). Each has two common characteristics that are of great significance. Some or all of the area on the bottom of the package is used for connecting to the circuit board, unlike the common perimeter packages, which waste all this "real estate" and thus crowd the pitch along the perimeter. The second feature is that virtually all use SMT. One may argue that the pin-grid array (PGA) is a feed-through device and thus is not surface mounted. This may be true in some cases, but most PGAs are plugged into a surface-mount device (SMD) package, and none of the pins protrude through the board. SMT devices do not pass through the circuit board, as do the older feed-through components. Certainly all the other styles with microspheres, solder balls, studs, columns, and any similar shape are SMDs. This is an important feature because it allows area array packages to fit into modern mainstream assembly processes. SMT has been the most common electronic component assembly process for many years, and it continues to gain more share because of high density and ease of automation. Fitting the SMT manufacturing infrastructure is important to growth and helps account for the popularity of area array packaging. Now let's examine the area array package concept more closely.

3.1 BASIC ELEMENTS OF A PACKAGE

3.1.1 Device

The area array package can be divided into simple elements. First, there is always at least one electronic device. There may be more than one integrated circuit (IC), and multichip packaging (MCP)

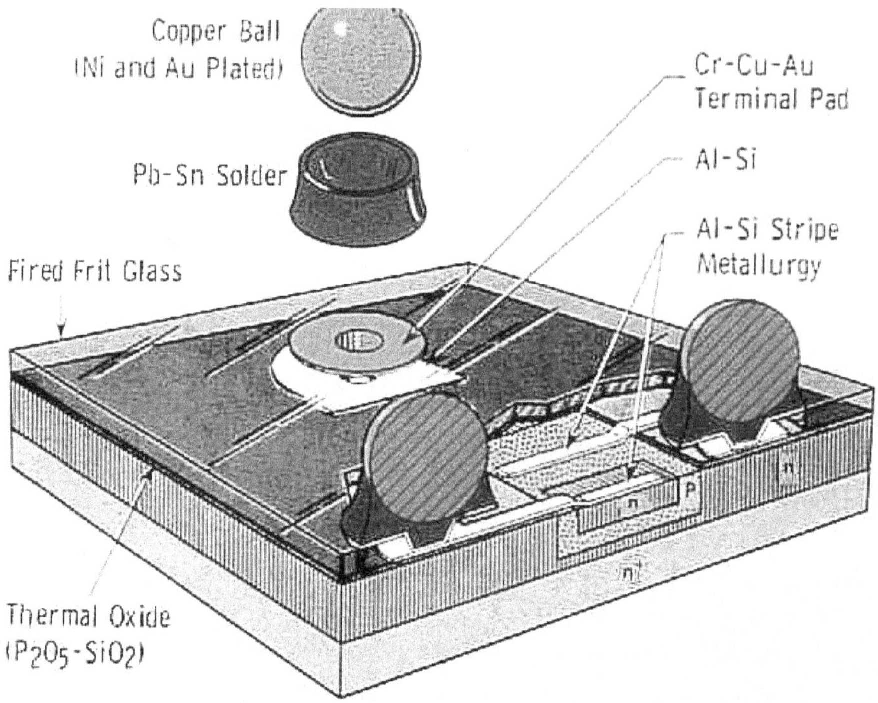

FIGURE 3.1 Solid logic technology (SLT). (*IBM Corp.*)

continues to gain strength. The device is typically a solid-state IC, but it also can be passive such as a resistor or capacitor array. In fact, tiny CSPs are now available to provide arrays of resistors, capacitors, diodes, light-emitting diodes (LEDs), and other optoelectronic (OE) systems. Computer chip packages often incorporate passive chips and sometimes memory.

3.1.2 Wiring or Routing

The next part of the system is some form of wiring structure that creates the pathway between the device (or devices) and the bottom of the package that ultimately will connect to the printed wiring board (PWB). There is almost always a geometric translation. The device dimensions are very small, and the wiring may route or fan out to a larger pitch that better matches the *practical* densities that can be produced using printed circuit technology. When the wiring pattern from the device spreads out and away, the design may be called a *fan-out pattern.* Conversely, the wiring may route conductors inward and under the device, and this is called a *fan-in pattern.* The fan-in pattern typically is used for very small packages and allows CSPs to be made. The common definition of a CSP is a packaged IC with a footprint not greater than 1.2 times the area of the die. An alternate but less common definition is a packaged chip where the total system is not greater than 1.5 times the volume of the die. These values are derived somewhat arbitrarily and have prompted terms such as *near-chip scale, near-chip size,* etc.

The wiring structure for area array can have several subelements. The package itself will have a pattern of electrical conductors typically produced by circuit board processes such as etching. Higher-density wiring is also being produced by additive and semiadditive methods where the conductors are produced by plating up. Other parts of the wiring structure may include wire bonds or whatever connection means is used to mate the package wiring to the chip. This is called the *first-*

level connection and will be covered later. There is also the ball, bump, stud, or column connection, and while this is part of the electrical pathway, it is not included as the wiring structure. The package wiring may be made of copper, gold, or a combination. It may be possible theoretically to use polymer thick film (PTF), but the limitations of this conductor medium have excluded its use so far.

A wiring layer also can be applied directly to the chip and sometimes is referred to as *rerouting*. Several companies offer commercial rerouting services, and these are usually combined with bumping of the chip pads to allow direct assembly. The resulting "package" may be called an FC or a CSP, and this adds to the confusion in distinguishing between these two competing designs. The process usually involves depositing a thin layer of dielectric over the bottom of the die, placing a pattern of conductors on the layer, and adding bumps. Most are wafer-level processes that produce all the packages simultaneously. The last process step is separation, called *singulation,* and is a cutting or wafer dicing method that typically uses a diamond saw.

3.1.3 Packaging Enclosure

The package must have some form of enclosure to protect the device, wiring structure, and chip-level interconnects. The enclosure can have other important uses, such as facilitating handling and board-level assembly, heat management, a surface for nomenclature, and other uses. The most common enclosure is made using polymer encapsulants, especially epoxies. Both liquids and solids are used, such as polymers, liquid encapsulant dispensing, and package molding. Some of the CSPs use special materials such as elastomers and chip coats. Although plastic packages are most common, ceramic and metal housings are also used, and these are covered elsewhere. See Chap. 8 for a detailed discussion of ceramic area array packages.

3.1.4 Board-Level Joining System

The materials and structures for connecting area array packages to circuit boards are where the most substantial differences will be found. While conventional perimeter packages use nonmelting metal leads, these are eliminated with area array. There are only two basic structures that can be used to populate underneath the package or die, and they are spheres (and bumps) and columns. Most joining structures are formed from metal spheres that are commonly referred to as balls to generate the common BGA term. Columns are also used because the greater standoff length provides special benefits in terms of joint fatigue resistance. Some FCs can have a rather flat or even mushroom-shaped bump that is produced by electroless plating without a resist to restrict the shape. However, a solder sphere or semisphere is produced on top of the structure by reflowing solder paste. Alternatively, conductive adhesive paste or film can make the connection.

3.2 OVERVIEW OF TYPES OF AREA ARRAY PACKAGES

Our definition in this book of area array packaging requires the existence of some area connections or the ability to produce them. In other words, the package technology must allow a board-level connection structure, such as bumps, to be produced within the substrate area, not just along the outer edge. Some area array layouts may only populate the outer row so that the package appears to be only a perimeter type. However, this can still meet the area array package classification if a second row of connectors can be produced. In all cases, the area array connection structure emanates from the bottom of the package, not from the sides like many perimeter designs.

The BGA is the most common type of area array package, and it offers a balance of features with reasonably good economics. The features include low profile, light weight, excellent reliability, and a quick time to market with many designs. Common ball pitches range from 1.0 to 1.50 mm. The package can accommodate over 1000 I/Os. Figure 3.2 shows a typical structure, but it should be noted that many other designs and enhancements are possible. CSPs and FCs can have much closer pitches, and the number continues to shrink.

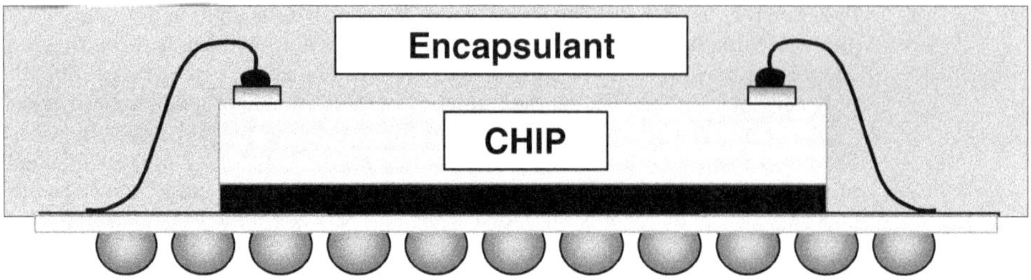

FIGURE 3.2 BGA cross section.

Multichip BGA packages became popular beginning in the late 1990s. They are the successor to the multichip modules (MCMs), and many people will say that the highly touted MCM was simply waiting for the area array package to be implemented. This may be correct, and the term *multichip module BGA* (MCM-BGA) is also used. The leading packaging foundries offer over a dozen multichip BGA styles. This package provides all the same advantages as the single-chip BGA but in addition provides simpler board layout and a reduced number of components to be assembled and often yields higher performance because of the closer proximity of the chips in package. The same Joint Electron Device Engineering Council (JEDEC) standard footprint outlines can be used that are designated for single-chip BGAs. Chips can be placed adjacent to each other in the same plane or stacked in 3D fashion (see Chap. 4).

Several companies offer thermally enhanced BGAs. A metal heat spreader or other thermally conductive material interface for coupling to a heat sink or cooler generally is incorporated. The heat spreader is nearly always placed in contact with the die for optimal heat transfer. Designs typically place the metal heat spreader on top of the package so that air can be blown across or a cooling device can be attached. The package carrier can be rigid organic, flex (tape), metal, or ceramic. However, the thin, flexible circuit–based packages can provide special thermal features. Heat also can be removed through the bottom of the package using a thermal block. Bumps, spheres, and columns also can serve as heat pipes to the PWB, and extra ones can be added even if they are not used for signal, power, or ground connections.

Tape- or flex-based BGAs are another important package, especially where enhanced thermal performance is desired. These packages are referred to as TBGAs, FBGAs, and fleX-BGAs (registered trademark of Amkor). The thin flex substrate offers very little thermal resistance. While all these BGAs are made with carriers that are tiny flexible circuits, the tape terminology has come about because of the similarity with tape automated bonding (TAB). TABs, also called *tape carrier packages* (TCPs), are really flexible circuits but without any balls or microspheres for attachment to the wiring board, and in fact, they are not even area array packages. The board-level bonding is done by soldering the perimeter leads with a heated bar. However, flex area array package designs retain the TAB inner lead bonding (ILB) format but replace the troublesome outer lead bonding (OLB) structure with area array bumps, balls, or spheres. The thinness of the flex dielectric makes it rather easy to fabricate vias through the bottom of the carrier that can accommodate attachment of solder spheres or some other interconnect structures. While TAB packages were limited particularly because they required special bonders, TBGAs have gained popularity because they are SMT components.

Several CSP designs based on flexible carriers are also available. The best known is the Tessera μBGA that is licensed to a number of companies now in high-volume production. The design is somewhat unique because the circuit pattern is fanned in to produce a nearly die-sized package. The connection to the chip uses an S-shaped lead (viewed as a cross-sectional) that is usually made from gold to avoid bond failure due to thermomechanical fatigue. These packages offer all the advantages mentioned earlier but have even a smaller footprint and lower profile. Electrical performance is considered to be modestly better than that of the larger BGAs because of shorter signal path lengths. This package is covered in Chap. 5, "Compliant IC Packages."

The flip chip BGA (FC-BGA), also called a *flip-chip-in-package* (FCIP), is an increasingly important package. It has the lowest profile and excellent electrical performance. A flip chip, or direct chip attach (DCA), rather than wire bonding is used to connect the chip to the carrier. The FC-BGA is especially useful with high lead counts. At some point between 600 and 800 I/Os, the flip chip economics become better than those of wire bonding. However, the exact number depends on many factors, and it is also a moving target. Continuous improvements in wire-bonding equipment move the crossover point to a higher number each year. However, as the die is reduced in size by "die shrink," a semiconductor densification process, and the number of I/Os goes up, a point is reached where flip chip is the right way to go or maybe the only solution. In fact, several central processing units (CPUs), or computer chips, can only be assembled by flip chip techniques. There are just too many pads to be placed along the die's perimeter even when staggered double rows are used. We should keep in mind that the flip chip is the only established chip-level connection that offers area array. Wire bonding and TAB are only perimeter methods.

The last area array package is actually the first, and it is the *flip-chip-on-board* (FCOB). Rather than building an FCIP, the bare chip is assembled directly to the wiring board. There are advantages but significant limitations for FCOB. First, the wiring board must have a very high density pad layout. There is no intermediate chip carrier to fan out the pitch. The next problem is that the assembled flip chip nearly always must be underfilled. Since underfilling is not part of the standard SMT assembly process, a second line generally is set up. The underfill line adds capital cost, material expense, and extra time. And the underfilled chip may not be reworkable, although new reworkable underfills are starting to be offered. Many, if not most, assemblers thus prefer the FCIP because the packaging foundry does the underfilling and the package can be reworked like any other BGA. However, when the minimum size, lowest profile, and highest electrical performance are critical, then FCOB is the right package. We should be aware that FCOB is used extensively in the mainframe computer area where it first started. Other FCOB applications include automotive controllers and cellular phones. Even though CSPs are supposed to eliminate underfill, this has not always been the case for cellular phones because CSPs can be damaged when the phone is dropped. Since many CSPs are underfilled to make them drop-proof, the flip chip becomes attractive again because it is smaller than any CSP. Also, products that have rerouting on the chip may be called CSPs, but they are not really more than a rerouted flip chip. Progress in underfilling eventually will minimize the cost impact. See Chap. 6, "Flip Chip Technology," for a discussion on new underfill methods.

3.3 ADVANTAGES OF AREA ARRAY

3.3.1 Density

The area array package greatly increases density, and this is probably the most important driving force that has made it a popular package today. The total number of leads T_p or I/Os, that can be placed along the perimeter of a package is derived from $T = 4N - 4$, where there are N leads or I/Os in a row because there are four sides to the package, and we need to subtract 4 so that we do not count the corner I/Os more than once. However, the number of connections for a fully populated area is $T_a = N$. The total number goes up dramatically as N increases. Table 3.1 shows how the total number of I/Os increases geometrically with the number of balls per row compared with perimeter-only. Figure 3.3 compares this graphically. Figure 3.4 compares perimeter layout with area array. Note how dramatically the area array layout exceeds perimeter as the value of N rises to 10 and higher.

3.3.2 Thermal Management

Although many standard packages have enhanced heat-management capabilities, area array is better suited to this task. In fact, there are families of area array packages designed specifically for high

TABLE 3.1 Perimeter versus Area Array

N I/Os per row	T_p, total perimeter leads	T_a, total area bumps
2	4	4
4	12	16
6	20	36
8	28	64
10	36	100
12	44	144
14	52	196
16	60	256
18	68	324
20	76	400
30	116	900
40	156	1600
50	196	2500

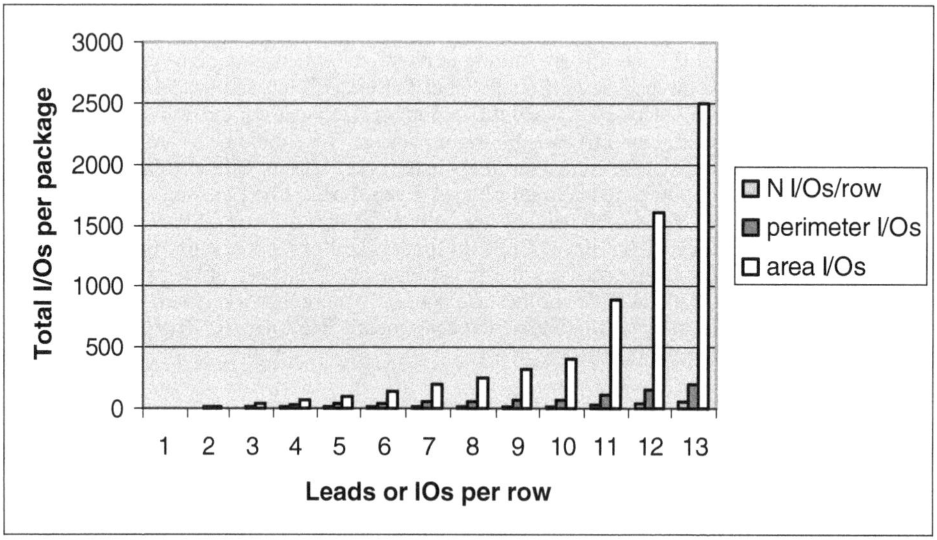

FIGURE 3.3 Perimeter versus area array (same pitch density).

heat dissipation. The flex-based BGA (also TBGA) has a metal heat spreader bonded to the thin flex carrier. Recall that heat transfer is inversely proportional to material thickness. This means that thin-flex, which can be as thin as 1 mil, is 10 times better at heat transfer than a 10-mil-thick rigid carrier with the same thermal conductivity. Since thermally enhanced flex polyimide laminate is available, the factor can be increased further. Ceramic, often an excellent thermal conductor, and highly conductive metal are also used for thermal packages with even higher efficiency, but at a cost. Figure 3.5 shows a thermally enhanced BGA.

3.3.3 Multiple Chips

Multichip BGAs, PGAs, and even CSPs have become available over the past several years, and their use continues to increase. While standard perimeter packages also can accommodate more

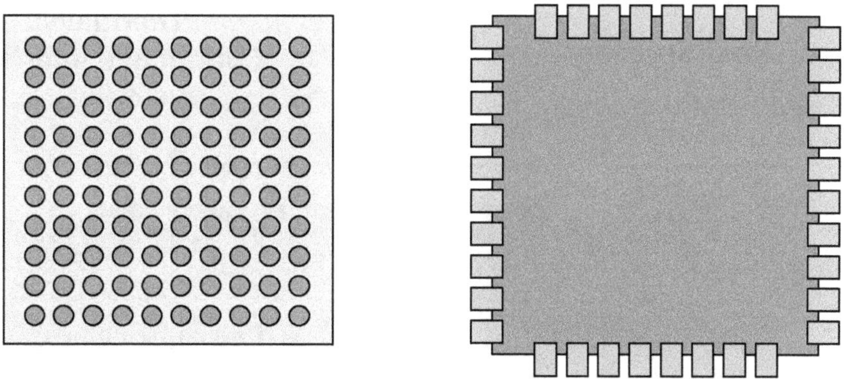

FIGURE 3.4 Perimeter versus area array layout.

than one chip, their use is more difficult. First, the use of multiple chips in a single package increases the number of package I/Os. The area array construction can accommodate many more I/Os than perimeter so that the package does not run out of capacity very quickly. Also, the ability to use FCIP with area array packaging allows closer packing than is practical with wire bonding. And finally, the ability to remove heat efficiently from the area array package allows higher density with less concern about overheating. Many people believe that that area array is the ideal form factor for multichip packages (MCPs), and the increasing popularity of this product segment supports this opinion.

3.3.4 Built-In Solder Source

Most area array packages are designed with the solder built into the package, thus eliminating the need to apply paste to the wiring board. Not only is the solder supplied with the package, the volume is much more precise than can be deposited onto the board by stenciling. However, flux generally is required, but this application process is much easier and less critical than for solder paste. Flux can applied to the bumps by an automatic coating process in which the pick-and-place machine dips the BGA into a reservoir of flux that is part of the equipment. The flux dipping process ensures that the flux is only where it is needed, on the bumps. Alternately, flux can be applied to the board, but the registration precision is low compared with solder's requirements. FCs and CSPs with eutectic solder also have integral solder, and this is even more beneficial with the typical fine pitch and small dimensions of these micropackages.

3.3.5 Self-Centering

When solder is melted to a liquid, the very high surface tension exerts considerable mechanical force. Surface tension attempts to reduce the surface area to a minimum. This is why solder forms spheres. When the solder ball of a BGA becomes molten, it wets the wiring board metal pad and then exerts force to minimize the joint area. A skewed or offset package usually will be forced into alignment. However, since area array packages have bumps on all four sides and often over most of their area, the alignment forces are stronger and more symmetric than on older perimeter-style SMDs. Work has shown that BGAs that are only half on the pad will orient. The centering works well, if not better, with CSPs and FCs because of their relatively lower weight.

3.3.6 High Assembly Yield

High assembly yields are due to some of the items listed previously. The precise amount of solder provided by the ball or bump typically precludes bridging or starved joints. The high level of

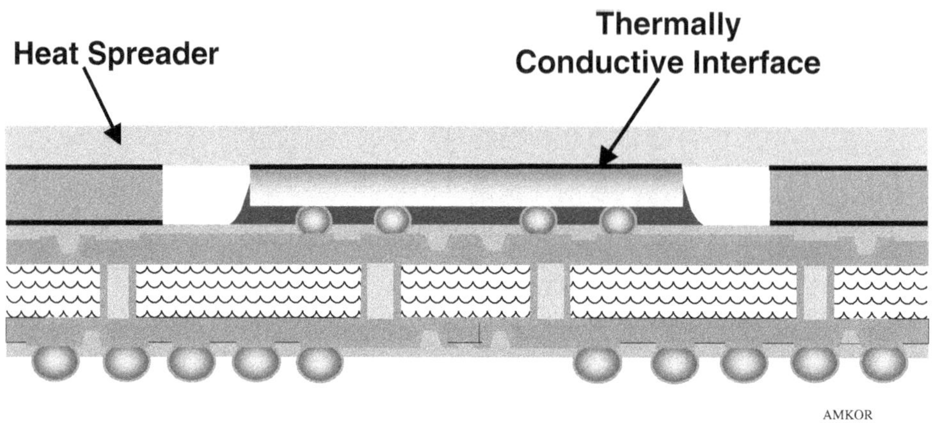

FIGURE 3.5 Thermal BGA.

self-orientation also is a big plus for yield. While one should attempt to place BGAs, CSPs, and FCs as accurately as possible, the self-alignment boost only adds to the high yield, where defects are often in the few parts per million range.

3.4 ISSUES WITH AREA ARRAY

3.4.1 Inspection

Many assemblers were frustrated in the early days of BGA assembly development because they could not see the solder joints under the package. We have relied on direct visual inspection for so long that it seems essential for quality, but this is not true. Difficulty of inspection still can be a problem today, although sideways-looking cameras can see further in but not perfectly. When accurate inspection is essential, real-time x-ray has become a preferred solution, albeit at a cost. Today, BGA joints can be inspected and analyzed quickly with digital x-rays with as much or more accuracy than vision systems for non-area array packages. Magnification, especially for CSPs and FCs, has been especially valuable for "seeing" small joints. Some people prefer acoustical imaging for FCs because even the underfill can be inspected. Underfill voids and even filler separation are well revealed.

3.4.2 Voids

Voids within solder balls (spheres) have been observed in BGAs before and after assembly by researchers and commercial assemblers. The source of the problem is not the solder sphere because the various manufacturing processes make it virtually impossible for voids to survive. All the solder sphere manufacturing processes go through a stage where the sphere is molten and the high surface energy of the metal quickly expels any gas or vapor. The common source of voids is volatiles from the flux either during the BGA balling process or during assembly to the PWB. Excessive flux has been the most common cause, but the wrong reflow profile contributes. Proper process and material control eliminate any voiding problems.

3.4.3 Cost

BGAs generally cost more than the equivalent perimeter SMD package. The BGA is generally more complex, and several conductor layers are common compared with a single metal lead frame for non-area array packages. While cost will continue to drop, the BGA may not appear to achieve parity. We need to keep in mind that there may not even be an alternate to the BGA, especially at higher I/O counts. The perimeter package becomes impossible to build at some point when I/Os increase and allowable footprint decreases. The BGA also may have enhancements such as heat management that may not be present in another style of package. Economics will continue to favor the use of perimeter-style packages when I/Os are low enough and area is not too restrictive. However, when certain density and performance criteria are important, then the BGA becomes the only choice. Make sure that you do not compare a BGA or CSP with a "phantom" package.

The FC is somewhat of a cost paradox. Since it is only a bare die with bumps, cost should be lower than for a standard package. Automotive electronics manufacturers such as Delphi-Delco claim that FCs are used to save money. Others complain about the high cost. The difference of opinion can relate to underfill and the added steps required. Underfill process equipment has been said to take up as much floor space as the assembly equipment. Underfilling remains an added step that imparts significant cost for many applications. However, continuous progress with underfill materials and processes will reduce and perhaps remove this cost adder. Aspects of underfilling are covered further in Chap. 6, "Flip Chip Technology."

3.4.4 Planarity

Area array packages must be reasonably planar for the solder balls to make contact with the wiring board and form joints. Once the balls collapse during solder reflow, they can accommodate significant noncoplanarity, however. There is no real problem for most ceramic and metal packages because these materials are very planar and stiff enough to resist warping stresses. The *flex-based BGA,* also called a *tape BGA,* typically is very planar because it is commonly bonded to a backer or stiffener such as metal. The *plastic BGA* (PBGA) can have planarity problems, however. Organic substrate, especially when thin, will warp out of plane if stressed. Warping stresses can be produced when encapsulant polymerizes. Nearly all polymers shrink slightly when they cure or polymerize. The shrinkage tends to warp the BGA organic carrier edges upward. This can happen with both liquid encapsulants and solid epoxy molding compounds (EMCs). The PBGA also may warp due to a thermomechanical mismatch between any of the components, but especially when the cured encapsulant and the carrier have different coefficients of thermal expansion (CTE values).

The early PBGAs had a significant warping problem due to excessive shrinkage of the liquid encapsulants that were used several years ago, as well as problems with substrate that was not yet optimized for this application. Today's encapsulants are considerably better because of improved polymers and higher filler levels. The higher the filler, the lower is the percentage of polymer that can shrink. Low-stress EMCs have come into use over the years and reduce warping even further. Substrate also has improved in terms of advanced polymers and better glass-reinforcement technology. However, a large PBGA made of thin substrate still can have warping problems.

3.4.5 Moisture Absorption

Moisture absorption by a package has at least two detrimental effects. Absorption by the dielectric can reduce electrical isolation and thus degrade the signal. A more serious problem can be the explosive evolution of water vapor in the form of steam on rapid heating. A PBGA can absorb enough moisture to delaminate or even crack the encapsulant during reflow soldering. The explosive evolution of steam by the package is known by the term *popcorning.* While the industry continues to reduce moisture absorption of BGA materials, the fix is to prebake the package just prior to solder reflow unless it has been kept dry in storage. Moisture in either the organic chip substrate, the die-attach adhesive, or the encapsulant can be a problem. Although some moisture absorption in the common epoxy encapsulants has been reduced, a different polymer with intrinsically lower moisture absorption may be the answer. At this point, moisture sensitivity is still considered one of the problem areas for PBGAs, including flex-based types. The polyimides used for flex can absorb up to nearly 3 percent moisture, making them one of the most sensitive package and circuit materials. In some cases, explosive evolution of water trapped in the polyimide film can blow metal conductors off. Predrying, while an annoyance, can be used for all packages. Flex circuitry is routinely predried before components are assembled.

3.4.6 Rework

BGAs are more complex to rework because the lead frame is the solder ball. While reworking procedures are well established and equipment is available, the process is not as simple as for perimeter packages with nonmelting connectors.

3.5 FIRST-LEVEL INTERCONNECT (CHIP TO CARRIER)

3.5.1 Wire Bonding

Wire bonding remains the most popular first-level connection process for BGAs and many CSPs. The process allows the wire-bonding machine to be programmed so that one BGA substrate design may accommodate many different chips of different sizes. Wire bonding continues to be an efficient process for all but the highest I/O types of chips or those dies with an area array pattern. Wire bonding over an active area (inside from the die's perimeter) is considered potentially destructive and not recommended even if it were practical. Continual rate improvements for die bonders allow dies with 600 I/Os to be bonded economically. The exact crossover point for moving to another first-level process, such as flip chip, continues to be a moving target as bonder rates increase. However, the point for moving away from wire bonding is probably between 600 and 800 I/Os. Wire bonding also increases the signal path length and therefore can limit the maximum external clock rate (frequency) of the chip. About 90 percent of all packages are still wire bonded, down from almost 100 percent a few years ago. The slow move away from wire bonding will continue, but it will remain a valuable process far into the foreseeable future.

3.5.2 (TAB) Tape Automated Bonding

TAB inner leads consist of very small cantilevered metal beams suspended over an opening (window) in the dielectric film of the chip carrier. The leads are patterned to correspond to the bond pads of the chip to be connected. The tiny leads typically are made of copper with a gold finish over a nickel-plated barrier. Some CSPs, like the Tessera μBGA, use pure-gold conductors. The connection is made by accurately positioning these inner leads over the chip pads and applying heat and pressure with a TAB inner lead bonder. The chip pads may be gold plated or even gold bumped to facilitate the bonding process. The pure-gold leads of the μBGA can be bonded to common aluminum pads, but TAB processing generally requires a gold-to-gold junction formation. One can view TAB bonding as a form of wire bonding where the "wires" are built into the carrier. The interconnect bonds can be formed one at a time or as an entire row (gang bonding). In some cases, all bonds are made simultaneously, making TAB bonding much faster than wire bonding. The TAB carrier is actually a flex circuit, but it often is made with thinner material and higher-precision circuitry processes. Standard TAB has a conductor pattern that fans outward and terminates as a wire pattern that can be soldered to the wiring board for second-level connections. However, this format is not area array, and the common TAB design must be modified. This will be covered under second-level, or package-to-board, connections later. Figure 3.6 shows the TAB inner lead and a "ball" outer lead that is commonly seen in many flex-based area array packages.

3.5.3 Flip Chip (FC)

FCIP generally is considered the first-level ultimate connection because the highest density can be achieved and the path length is shortest so that optimal electrical characteristics are achieved. The FC has bumps over the bond pads that are used for the connection. These bumps may be made of solder so that the bump itself is the source or of a nonfusible metal or alloy. All the FC bonds are made simultaneously, making this the fastest first-level connection method. FC is also the only area array first-level process, making it the densest format. The total number of I/Os possible for a single chip probably can exceed 10,000. IBM has built actual products with over 7000 bumps. Both organic and

FIGURE 3.6 TAB with BGA format.

ceramic carriers are used with FC. CPUs, better known as *computer chips,* are most commonly assembled as FCIP, and I/Os range from 1000 to 3000, although this number continues to climb. This technology is well covered in Chap. 6, "Flip Chip Technology."

3.6 PROTECTION METHODS

3.6.1 Epoxy Transfer Molding

Epoxy molding compound (EMC) is the most common encapsulant used to protect packages, and this is true of BGAs as well as other styles. Individual packages, strips, or even full arrays of unsingulated carriers can be molded.

3.6.2 Liquid Encapsulation

Liquid epoxies and other polymers can be applied to encapsulate an area array package using a needle dispenser. Automated dispensers are used widely that can handle the entire process, which may require several steps. A common method called *dam and fill* involves needle dispensing a retaining bead, or "dam," of viscous encapsulant around the package perimeter. Fluid "fill" encapsulant is next deposited within the boundary to cover the chip and interconnect structure. The assembly is then heated to harden both materials. Liquid encapsulants are used more commonly for prototyping, for shorter production runs, and in situations where fast time to market and minimum production tooling are important criteria. Liquids are used almost exclusively for cavity-type packages, where the die is located in a recess in the bottom of the carrier, because molding is not practical.

3.6.3 Lid Seal

Some metal and ceramic packages may be sealed with a cover or lid that is bonded over the opening once the chip is attached and connected. Metal lids can be soldered, welded, or bonded adhesively, although the latter method does not give an airtight hermetic package. Ceramic packages and lids also can be soldered if the appropriate metalization is applied. They also can be bonded adhesively. In all cases, an adhesive lid seal does not provide an airtight hermetic enclosure because air slowly leaks through the polymer. Continuing improvement in polymer seal materials may provide near-hermetic or quasi-hermetic packaging that is good enough for most situations that need a low-gas environment.

3.6.4 None?

Some people view the FCIP when not covered with encapsulant as unpackaged. However, the active side of the die faces downward, and the underfill serves as the protective encapsulant that seals the interconnect structure. Very few, if any, area array packages are unprotected. Chips are not only sensitive to the environment, the sensitivity also has been increasing as denser, more sophisticated products have been introduced. A few CSPs may be protected by inorganic "encapsulant" instead of polymers. The early IBM STL is an example.

3.7 SECOND-LEVEL INTERCONNECT (PACKAGE TO PWB)

3.7.1 Pins: Pin-Grid Array (PGA)

The PGA uses a mechanical connection method by incorporating an array of stiff metal pins into the package substrate that can plug into a socket on the PWB. The pins are usually a base metal with gold plating to ensure a good, unoxidized system for mating. The PBA body can be organic, ceramic, or metal. Ceramic has become popular because of strength, low thermal expansion, high elasticity modulus, good planarity, high thermal conductivity, exceptional heat resistance, and excellent electrical insulation. Since the second-level assembly process is mechanical, easy field replacement is possible, making this a good package for chips that need to be upgraded, such as CPUs. This type of package remains popular, and FCIP is common for CPUs. The higher cost, however, makes ceramic a target for organic substrate.

3.7.2 Eutectic Solder Spheres

Most BGAs, CSPs, and FCs use eutectic solder for second-level connections. Eutectic solders, by definition, have a melting point (liquidus) and freezing point (solidus) that are equal. Exceptions are high-melting-point alloys that require the addition of solder to make the connection to the board. Eutectic solder typically is supplied as the BGA ball or FC bump, and no additional joining material needs to be added.

3.7.3 Nonfusing Metal Spheres, Bumps, or Columns

Nonfusible means that the metal or alloy does not melt under typical solder reflow conditions. Nickel, gold, copper, high-lead solders, bumps, balls, and studs are all used. Solder or conductive adhesives are used to form the joint between the package metal and the wiring board. Advantages for nonfusible systems include an easier test interface and a predictable stand-off height, or gap, which can be valuable for FCs. The gap for FCs has a great influence on joint fatigue, and underfill cannot flow under too small a space.

3.7.4 Conductive Adhesives

Conductive adhesive bumps are used for FCs but are not common on other area array packages. Work to apply them to BGAs did not show any real advantages over solder balls.

Isotropic Balls/Bumps. Isotropic conductive adhesives (ICAs) are typified by silver-filled epoxies similar to those used for die attachment. The adhesives can be applied to the chip as a paste by stencil printing, preferably at wafer level, and then hardened thermally. The polymer-bumped chips are then bonded to PWBs by adding more paste than is usually printed onto the PWB. The bumped FC is placed into the paste and then thermally cured. Underfill is required. Some versions involve applying a nonconductive film of underfill to the wafer or PWB. The adhesive bumps are forced through the film by heating and applying force.

Anisotropic Films. Films or pastes can be applied to FCs to provide an interconnect using anisotropic conductive adhesives (ACAs). The film or paste contains a dispersion of tiny conductive particles, usually spheres, that will create a path between the chip pad and wiring board when force is applied. The conductor level is too low for paths in the *xy* plane to be produced, and it acts like an insulator. Only particles trapped between opposing chip and circuit pads form electrical paths. No underfill is needed because the adhesive layer serves this purpose.

3.8 THE PLASTIC BALL GRID ARRAY (PBGA)

3.8.1 Die-Up

The simplest area array package has a die-up (cavity-up) design (Fig. 3.7). The chip is bonded to the top of the carrier substrate opposite to the ball or stud connection side. The chip can be wire bonded or attached directly in an FC format. The chip and interconnect are encapsulated by liquid or EMC in the case of wire bonding. An FC may be underfilled but also overcoated with encapsulant. There are molding processes that underfill and overcoat simultaneously.

3.8.2 Die-Down

The die-down (cavity-down) design cannot have a full array of bumps because space must be reserved for the die cavity. The package must be deep enough that a die cavity can accommodate one or more chips as well as the interconnect structure such as wire bonds. The package can be organic, ceramic, or metal. Once the chip is bonded and connected, the package must be sealed with a lid or filled with liquid encapsulant to protect the die. Figure 3.8 shows a typical die-down format.

3.9 METAL PACKAGES

Metal area array packages are designed with insulation to isolate conductors. The chip is placed and bonded within the cavity and then connected electrically.

3.9.1 Hermetic

A truly hermetic seal requires that a metal lid be sealed metallurgically, with solder or by welding.

3.9.2 Encapsulated

Filling with polymer encapsulant can protect the die and connection structure within the metal cavity. This is a simpler and lower-cost process, but gas molecules will diffuse slowly through the plastic. This is not a problem for most dies.

FIGURE 3.7 Die-up BGA.

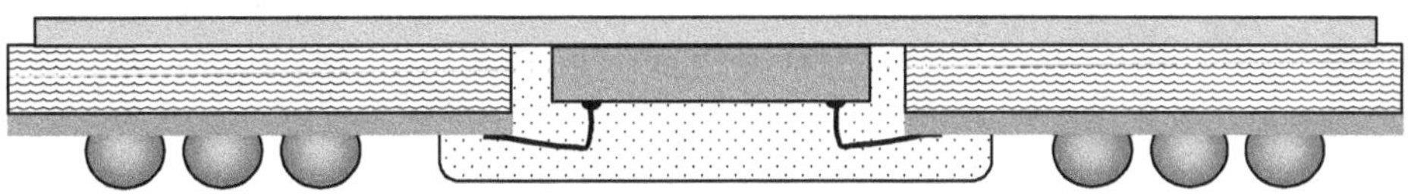

FIGURE 3.8 Die-down BGA.

3.10 CERAMIC

Ceramic BGAs and PGAs remain popular for high-lead-count chips such as computer chips (see Chap. 9). The high degree of planarity, excellent thermal stability, and good thermal conductivity override the higher cost factor. Ceramic modules can be used as a carrier for FCs, sealed with a lid to form hermetic package, or used with underfill. The most popular format for computer chips is a carrier for an FC. The board-level connection can be mechanical pins, solder balls, or solder columns.

3.11 CSP PRODUCTS

CSPs come in all shapes and configurations. The simplest use an organic carrier platform with solder balls on the bottom. The chip is bonded to the top of the carrier and connected with short-path and low-loop wire bonding. The only difference between a BGA and a CSP is really one of relative dimensions of die versus package footprints.

3.12 FLIP CHIP: IS IT A TRUE PACKAGE?

There is no doubt that the FCIP is the genuine article, a real package. What about FCOB? The consensus is that an underfilled FC is not reworkable, at least in the common meaning of the term, and therefore is not a package. Most, but not all, packaging authorities suggest that a package should be removable. The ideal package is *reworkable,* meaning that it can be used again on a new wiring board or be repositioned if the fault was with the assembly. This is especially important for computer chips, which can have much more value than a PWB.

3.13 RELIABILITY

The area array package is one of the most reliable designs. The FC on ceramic substrate used in essentially all the IBM mainframe computers probably holds the record for reliability. The ability to distribute signal and power more evenly and to remove heat more efficiently gives area array a reliability advantage. Early BGAs had some reliability problems as the technology was just emerging. Improved designs, optimized materials, and a better understanding of the system have eliminated these initial reliability issues.

3.14 FUTURE EXPECTATIONS

The various area array designs can be expected to gain more market share as the desire for higher performance and larger-I/O-count products increases. While many people had hoped for a consolidation of packaging types, the trend will be for a continued proliferation of types. We can expect more CSP and BGA packages to be developed as the need for unique packaging solutions increases, especially since new microelectromechanical systems (MEMs) and advanced photonics products add special requirements that may be best handled with area array designs.

CHAPTER 4
STACKED/3D PACKAGES

Lee Smith
Amkor Technology, Inc.

4.0 INTRODUCTION

Industry experts believe that the fourth wave of semiconductor package innovation is being driven by challenging new electronic applications. The term *application-specific packaging* is being used to characterize the new microelectronic innovations associated with this era of electronic system convergence. This fourth wave is enabling further silicon integration at the system level by allowing a mix of semiconductor technologies to be integrated at the package level to address specific subsystem density requirements. This chapter will look at both the mobile phone market as the new "killer" application driving the magnitude of this fourth wave and the stacked or three-dimensional (3D) package as one of the most promising new application-specific packaging innovations available. Figure 4.1 provides a definition of application-specific packaging and describes the scope of requirements being addressed through multichip integration at the package level.

4.1 THE FOURTH WAVE OF PACKAGING INNOVATION

Figure 4.2 is Prismark Partners' look at the past and projected waves of packaging innovation over a 40-year horizon.[1] The first wave of packaging was through-hole components driven by the large motherboard assemblies associated with the telephone and computer industries that dominated the late 1970s through early 1980s. When the wave of new low-cost consumer electronics began to flow in the middle to late 1980s, innovations in packaging and assembly were needed to provide the automation required to deliver the high volumes and low costs needed to satisfy strong consumer appliance demand. Thus the surface-mount component was developed and saw rapid adoption in both consumer and emerging personal computer (PC) applications. As packaging pin counts and package sizes continued to increase to support the device advances associated with very large scale integration (VLSI), the perimeter leaded surface-mount technology (SMT) component began to be a limiting factor for automatic assembly and miniaturization due to the sensitive nature of the fanned-out fine-pitch surface-mount leads. Thus a new innovation that distributed the contacts across the entire area of the component was born in the 1990s to support the increasing input-output (I/O) demands of the PC wave. This third wave was called *area array* and was characterized by the plastic ball grid array (PBGA). It provided significant size reduction and SMT throughput increases due to the benefits of the self-aligning solder balls evenly distributed under the package. Now that the mobile phone handset has become the "killer" application of the new millenium, a fourth wave of packaging innovation is required to allow handsets to continue to miniaturize while increasing in

Definition:

- A package developed or customized to meet the specific requirements of an electronic subsystem

- Scope of requirements:
  - Reliability specific to that application (1st & 2nd Level)
    - 1st or component level testing defines the reliability of the package over the range of assembly temperatures and use environments expected over the life of the product
    - 2nd or board level testing defines the reliability of the package solder joint to mother board interface over the products life
  - Performance specific to the application and devices or functions integrated within the Package
    - Form Factor, Electrical, Thermal, Mechanical, Optical
  - Designed for optimization of the total cost required for the application
    - Standards, Infrastructure, Time to Market

FIGURE 4.1 Application-specific packaging.

functional performance. In addition, the growing density gap between the semiconductor industry (which has remained on track with Moore's law for doubling every 18 months) and the slower-moving printed wiring board (PWB) industry has required that the integrated chip (IC) package provide the critical space transformer—between the submicron density of the IC industry and the much courser submillimeter density of the PWB world. One of the most effective ways for the IC packaging industry to keep pace or scale with Moore's law without outpacing the PWB electronic platform's ability to handle more silicon integration is through 3D packaging—volumetric solutions. The mobile phone market was a strong early adopter of chip-scale packages (CSPS) and also has been a leader in seeing the benefits 3D packaging provides for increasing memory capacities in shrinking handsets.

To help put Fig. 4.2 in perspective, in 1999, over 68 billion ICs were produced worldwide, 94 percent of these housed within a package. Prismark projects over 113 billion ICs in 2004; assuming a conservative estimate of 15 percent unit growth through the year 2020, then the 3D packaging share of 11 percent can total to well over 70 billion units that employ stacking technologies. Thus, even if this forecast is off by a few percentage points, the market for 3D packaging is expected to be huge going forward.

To achieve continuous size, weight, and cost reductions, wireless handset manufacturers and their semiconductor suppliers have been developing new semiconductor devices and packaging technologies that enable higher levels of functional integration at the smallest, most cost-effective level, which is through silicon (Si) integration. The higher levels of Si integration for baseband functions have developed the quickest due the system-level integration nature of application-specific ICs (ASICs) and the common CMOS processing for baseband functions. The memory and radiofrequency (rf) functional integration for cellphones is currently being accomplished through application-specific packaging advancements defined as *system-in-a-package* (SiP). The emergence of SiP

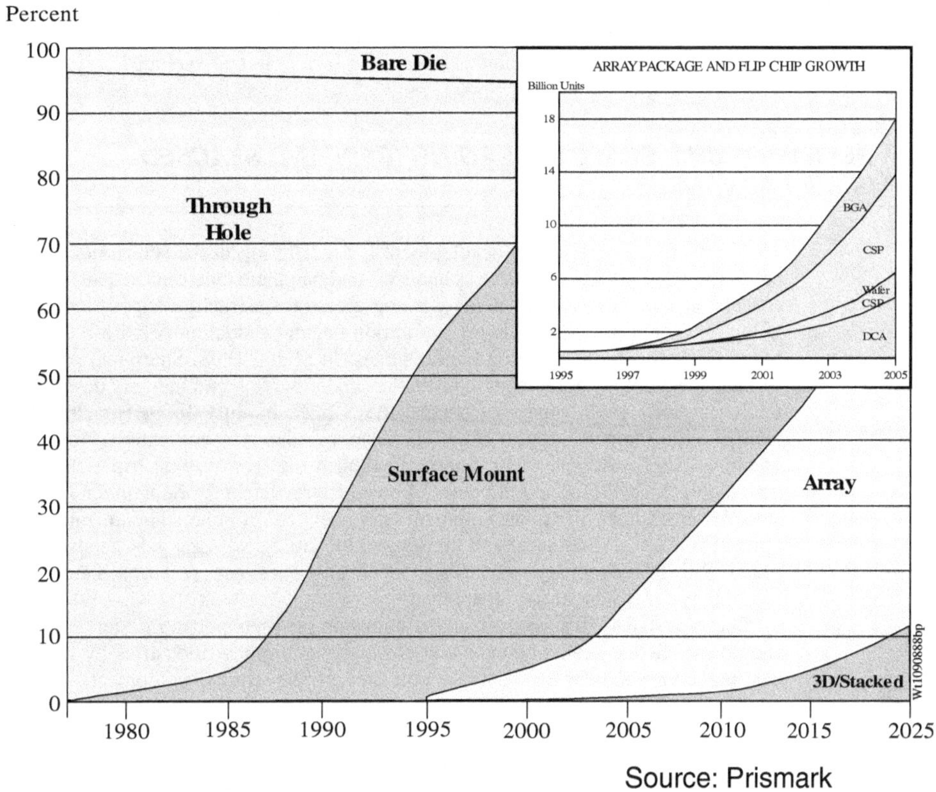

FIGURE 4.2 Waves of electronic packaging.

is due to the complications of integrating different IC technologies or densities in a single chip or system-on-a-chip (SoC) solution. This chapter will take a look at advancements in thin die stacking delivered through 3D packaging and more specifically stacked chip-scale packaging (S-CSP) technology, which is a major segment of the SiP movement and application-specific packaging trend. To avoid the confusion associated with the terms *application-specific packaging, system-in-a-package,* and *3D packaging,* consider the following guidelines:

- Application-specific packaging defines the *trends* that are driving a wide range of package advancements to meet the requirements of a specific segment or segments of device and application cost/performance requirements.
- System-in-a-package (SiP) defines the technical approach of integrating mixed semiconductor technologies at the package level where multiple ICs and their associated discrete and passive components are integrated to form a subsystem functional block that is fully characterized and tailored to meet that application's performance requirements.
- 3D packaging is a subsegment of the SiP movement where die or package stacking is employed to provide higher levels of silicon efficiency with respect to the package area by using the z dimension for volumetric density solutions.

3D packaging allows further system integration through stacking of diverse device technologies, including Flash, static random access memory (SRAM), and dynamic random access memory

(DRAM) memory blocks, as well as baseband, mixed signal, analog, and logic functions. Stacked chip package integration will be critical to allow the pace of handset size, weight, and cost reductions to continue, as well as new handset form factors to be developed.

4.2 REVIEW OF CHIP-SCALE PACKAGING (CSP) AND STACKED CSP (S-CSP) ADVANCES

The pace of CSP technology development is accelerating in the semiconductor industry, driven by broad adoption of CSPs in wireless handsets and handheld electronics. The following milestones are cornerstones of the foundation for a strong global infrastructure supporting CSP and S-CSP technologies. In August 1996, Sharp Corporation began production of wire-bonded, flex circuit, tape-based CSPs for initial adoption in cellphones. In March 1998, Sharp and Amkor Technology, Inc., formed a technology agreement to promote standardization and industry development for tape-based CSPs. In April 1998, Sharp introduced the S-CSP; by July 1999, Hitachi, Mitsubishi, and Intel teamed with Sharp to support S-CSP standards (industry specifications) for Flash and SRAM die combinations. In August 1999, Sharp announced the world's first triple-chip S-CSP development (triple-chip S-CSP production plans followed a year later). In September 1999, Sharp and Amkor signed a cross-licensing codevelopment agreement to promote standardization and industry development for S-CSP laminate- and tape-based technologies. Amkor is licensed to use Sharp's tape-based S-CSP technology, and Sharp is licensed to use Amkor's laminate-based ChipArray technology for S-CSP and CSP applications.

This type of S-CSP agreement and relationship between industry leaders such as Sharp and Amkor should benefit the semiconductor and electronic equipment industries by teaming to drive S-CSP cost reductions and ensure reliable supply and infrastructure development.

4.3 HANDSET FUNCTIONAL SYSTEM INTEGRATION

The continuous size, weight, and cost reduction trends for handsets would not be possible without the tremendous capabilities of the industry supply base to respond to new challenges and the insatiable levels of wireless consumer demand. The mobile phone broke the price elasticity supply-demand model and created a new model of the greatly discounted or free system carried by the wireless communications provider through the service contract. This delivery model contributed to the tremendous rates of wireless adoption and phone replacement rates as better handsets and services emerged each year.

Figure 4.3 provides a multiaxis graph of the reduction trends that have helped drive tremendous worldwide handset adoption. This graph provides a macro view of the rates of change over the past 5 to 10 years.[2]

1. 15 percent annual reduction in weight.
2. 14 percent annual reduction in PWB area.
3. 25 percent annual reduction in cost/price.
4. 60 percent annual increase in unit volume.

However, to understand the roots (enablers) of this ongoing level of change, an understanding of the functional blocks and their associated levels of Si integration and high-density interconnection (HDI) is required. Figure 4.4 summarizes the functional block diagram of a typical high-volume mobile phone, single-mode Global System for Mobil Communications (GSM) handset. Up through 1998, high levels of Si integration in the baseband subsystem have been key in enabling handset advancements. The three ASICs in the baseband subsystem of Fig. 4.4 illustrate this strong baseband integration.[3]

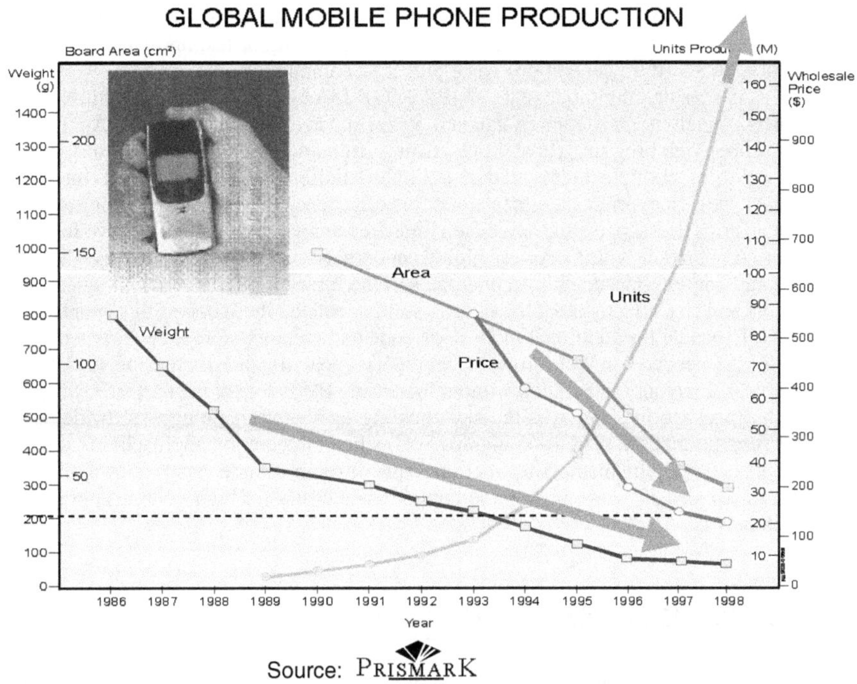

FIGURE 4.3 Global mobile phone production trends.

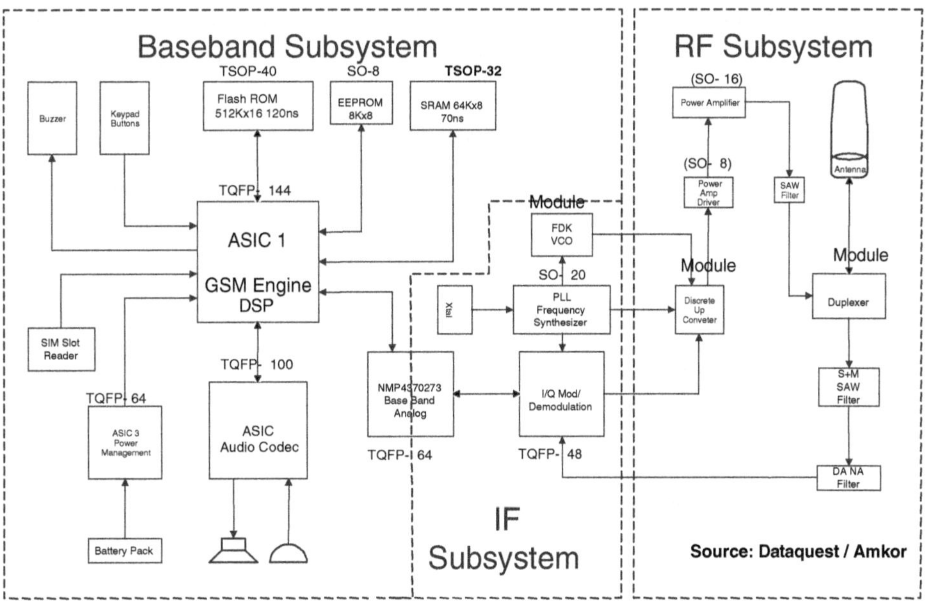

FIGURE 4.4 A wireless handset is a complex system in a few cubic inches.

The majority of 1999 and 2000 model handsets include multimode and multiband functions for broader regional service connectivity and are planning or testing features that offer generation 2.5 through 3G (third generation) voice plus data communications.

As a result, multimode (e.g., AMPS + TDMA or GSM + CDMA), multiband (e.g., 900 and 1800 MHz) handsets are becoming standard, with I-mode (Japan) and Wireless Application Protocol (WAP, Europe) Web browsing capabilities gaining strong adoption. To achieve size and cost-effective integration of multiple radios within a single handset, multichip package integration (system-in-a-package) is the current industry answer. Radiofrequency (rf) integration is being accomplished in package platforms that support the associated high-frequency materials and passive integration requirements in combination with GaAs- or high-frequency Si-based semiconductors. (Multichip rf packaging could employ stacked chip technologies in the future when the rf shielding capability is made more cost and size effective, but the rf subsystem is outside the scope of this chapter.)

To handle the additional multimode code and memory storage requirements, along with the data storage needed for Web browsing, memory capacities are increasing rapidly. To achieve higher levels of memory integration without increasing PWB area or handset size, weight, and cost, stacked chip package integration is the key enabling technology gaining worldwide adoption. Despite the strong promotion, system-on-a-chip (SoC) silicon integration approaches that require large memory blocks (or require memory technologies that differ in volatile versus nonvolatile architecture and feature size density) have proven too complex and costly for high-volume applications such as mobile phones.

4.4 CSP TO S-CSP ADOPTION IN HANDSETS

The mainstream 1998 handsets (as represented in Fig. 4.4) used traditional perimeter leaded IC packaging technologies based on copper lead frames. Small-form-factor "personal handy phones" available from Japanese suppliers in 1998 and mainstream handsets for 1999 adopted CSP technologies to reduce size and weight. The initial devices and CSPs adopters were

1. Tessera's widely licensed μBGA (TAB bonding) for Flash and SRAM packaging.
2. Wirebond on flexible printed circuits (polyimide tape) or rigid laminate-based CSPs/FBGAs (fine-pitch plastic ball grid arrays) for digital signal processors (DSPs) and SRAM and electronically erasable programmed read only memory (EEPROM) devices. Leading tape-based CSP packages include Texas Instruments' MicroStar BGA and Amkor's *fle*XBGA. Motorola's MAP (mold array package) and Amkor's ChipArray were leaders in the wirebond on rigid substrate package families, with each employing gang mold and saw singulate process technologies for improved die to package edge design rules and reduced tooling/nonrecurring engineering (NRE) costs for each new package size.

These first two classes of CSPs became widespread in late 1999 and early 2000 handsets. Before moving forward to the S-CSP, it is important to help clarify the blurring definition of a CSP. Early on, a CSP was defined by Joint Electron Device Engineering Council (JEDEC) as a package only 20 percent larger than the die it contained. However, this definition did not really lend itself to standardization and infrastructure development. From the original equipment manufacturer (OEM) and surface mount technology (SMT) perspectives they did not want to see a motherboard change every time a CSP die shrunk, thereby requiring a new footprint or package outline. The tray and test socket suppliers had to wait for clear customer demand for each new CSP, creating a longer time-to-market gap for new CSPs than the rapidly moving wireless industry could allow. Also, every new minimalist package being developed wanted to be characterized as a CSP (due to the growing interest and market demand) while at the same time solving the disadvantages of true CSPs (by allowing a wider range of die sizes and shrinks to be compatible, without the need for a package change or costly requalification). In addition, the CSP definition had no application to the emerging multichip packages being adopted in handset and portable applications. Thus the distinction or lines between BGAs and CSPs really began to blur, making it hard for customers and market forecasters alike to characterize and segment for evaluation. Therefore, a more practical definition of the CSP was developed for area array packages. The new distinction was based on the associated solder ball pitch, with CSPs characterized as array packages with ball pitches of 0.8 mm and below and BGAs as ball pitches of 1.0 mm and above. This distinction has worked pretty well from the OEM down through the supply chain because now the distinction is more a factor of density rather than die-to-package size. A fine-pitch density segmentation also has application to the type of SMT assembly and PWB fabrication required to support the finer-pitch devices.

As soon as this new definition of CSP achieved broad acceptance, a new 3D minimalist packaging approach emerged in order to continue meeting handset cost and form-factor reduction requirements for their increasing memory capacities. Now we see S-CSP technologies being adopted by nearly all major handset manufacturers for the memory blocks and baseband section. Figure 4.5 illustrates the space savings associated with integrating Flash plus SRAM into a S-CSP versus two separate FBGAs. For a 16-Mb Flash device and a 2-Mb SRAM (common in the initial transition from CSP to S-CSP for 2000 mainstream handsets), the associated PWB area savings is 60 mm^2.[4] Figure 4.6 provides a size and weight comparison for S-CSP versus the alternative Flash and SRAM single-chip and multichip packaging approaches.[5] Two separate thin small outline packages (TSOPs) provide the baseline (100 percent) for size and weight reduction comparisons. As the first column indicates, 16 Mb of Flash and 2 Mb of SRAM integrated in a S-CSP are about one-third the area and weight of two separate TSOPs common in 1998 model handsets. In other words, S-CSP provides a factor of 3 increase in packaging density. In addition to the preceding benefits for Flash + SRAM in an S-CSP, in some cases the EEPROM (data storage) block or function is being integrated within the base Flash chip or its architecture, allowing read while write (RWW) functionality.

These Flash device and packaging advancements allow handsets to integrate higher memory capacity in a single package versus three separate components, providing greater silicon efficiency through 3D packaging density and the associated cost reduction.

4.5 S-CSP GROWTH: STANDARDS AND INFRASTRUCTURE

The electronics and semiconductor markets saw a slowing in their respective torrid growth rates in the fourth quarter of 2000. Subscriber growth rates for wireless communications also have slowed in the largest markets, the United States and Europe, due to economic softening and high rates of wireless adoption (market penetration or saturation). Coupled with a decline in replacement or upgrade

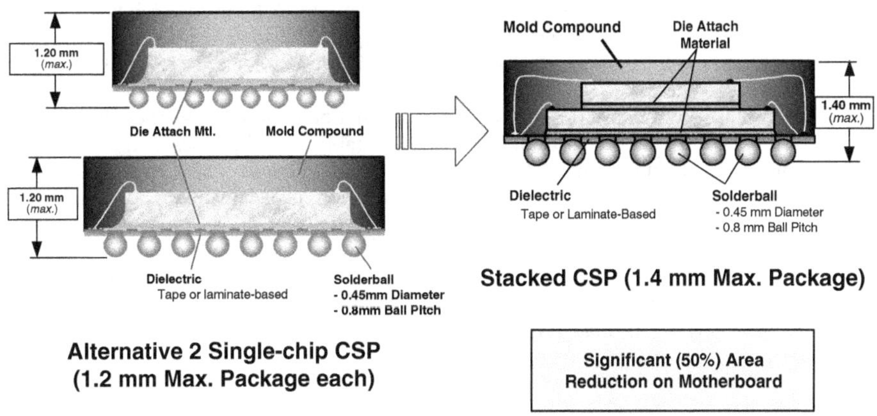

FIGURE 4.5 Flash plus SRAM integrated into S-CSP.

	Stacked CSP	TSOP x2	CSP x2	Stacked TSOP	MCM/Module
Structure (*Memory*)			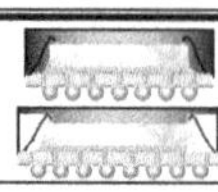		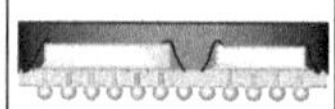
16Mb Flash (x8)	8x10mm 80mm^2 0.17g	10x14mm 140mm^2 0.30g	8x10mm 80mm^2 0.13g	10x14mm 140mm^2 0.32g	11x14mm 154mm^2 0.34g
2Mb SRAM (x8)		8x13.4mm 107.2mm^2 0.25g	6x8mm 60mm^2 0.08g		
Mounting Area (mm^2)	80	247.2	140	140	154
	32%	100%	57%	57%	62%
Weight (g)	0.17	0.55	0.21	0.32	0.34
	30%	100%	38%	58%	62%
Package Height (mm)	1.4max. (1.25 typ.)	1.2max. (1.10typ.)	1.2max. (1.05 typ.)	1.2max. (1.10 typ.)	1.6max. (1.40 typ.)

Highest Silicon Efficiency

FIGURE 4.6 S-CSP versus alternatives.

phones and services, this has offset the tremendous rise in I-mode adoption within Japan to downgrade worldwide cellphone consumption forecasts for 2001. Industry estimates vary, but on average, forecasters and handset suppliers report that around 420 million handsets were sold in 2000. Not all these sales ended up in the hands of consumers, with service providers and distribution channels accumulating excess inventory, further causing demand adjustments throughout the supply chain. Since Flash memory has been in tight supply through 2000, inventory adjustments have had less impact on the growing demand for S-CSP.

TechSearch International estimates that 100 million S-CSP products were consumed during 2000 almost exclusively from cellphone applications. Figure 4.7 provides TechSearch's recent forecast for cellphone growth, S-CSP consumption, and the rate of penetration into cellphones.[6] Based on the slow start in 2001, this forecast is seen as optimistic for cellphones over the next 4 years. However, the forecast is conservative for 500 million S-CSPs in 2004 due to the broad range of new 3D packaging applications from personal digital assistants (PDAs), wireless base station memory blocks, Flash memory cards, and Internet switch/router memory blocks being planned for next generation systems.

This enormous 400+ million unit consumption established wireless handsets as the overwhelming highest volume electronic system and establishes cellphones as the "killer" application for the new millennium. Further, the handset becomes the dominant communications platform and subscriber base for further integration of information management applications such as Internet access (Web browsing), which is at the center for the delivery of these *new data-centric wireless services,* as can be seen in Fig. 4.8*a,* along with the innovative new handset form factors being tested or developed in Japan, as can be seen in Fig. 4.8*b.*[7,8] The following innovative new handset form factors, features, functions, and services clearly require innovative new packaging solutions, including 3D packaging, to allow added functionality without increasing handset size and weight.

- Voice mail and text messaging
- Faxes
- E-mail

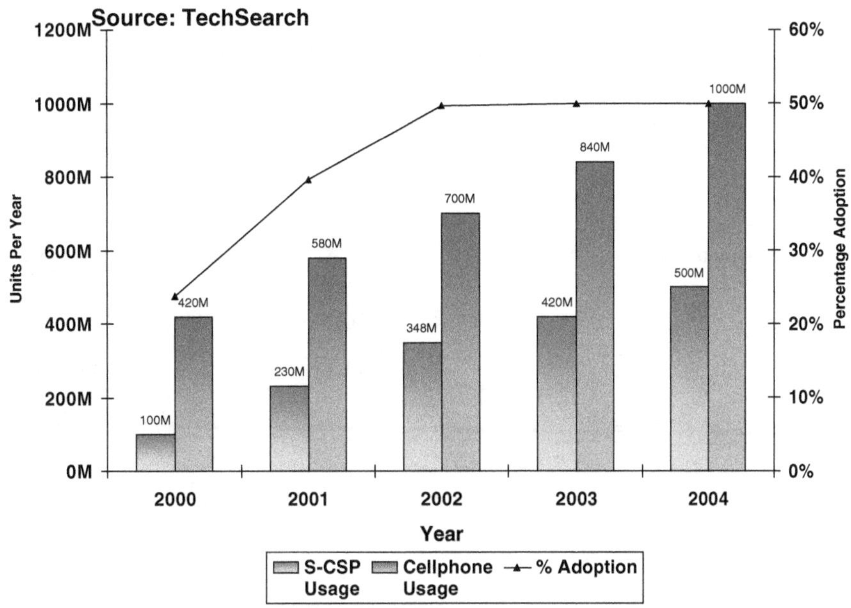

FIGURE 4.7 S-CSP market forecast.

- Internet browsing
- Maps
- Airline and other online reservations
- Telephone directories
- Over the air software (program) updates
- Stock quotes and transactions
- News clipping services
- GPS (global positioning system)

These data-centric applications will drive additional code and data-storage capacity requirements. Further, the availability of diverse services will cause application-specific segmentation of the handset platforms that must cost-effectively support their base voice communication services along with this range of additional new services. Some of the industry leaders are indicating there may be as many as five different handset segments for different levels of budget, service, features, and functions. This surely will drive the need for a robust industry infrastructure that meets not only today's need for high-volume availability of the latest semiconductor, packaging, microvia, and thin high-density substrate technologies but also tomorrow's needs for even higher levels of integration and density. Figure 4.9 provides an outlook for component and assembly densities expected for three major handset segments by 2004.

The following are some of the current items essential to ensure that the industry infrastructure is in place to support the expanding requirements of these additional cellphone segments along with their shrinking form-factor and time-to-market demands:

1. Second-source availability of Flash memory components for cellphones, since leading Flash suppliers continue to have different strategies, protocols, and pin-outs for their leading-edge designs. The handset makers drive their Flash suppliers for innovative products that solve their storage problems

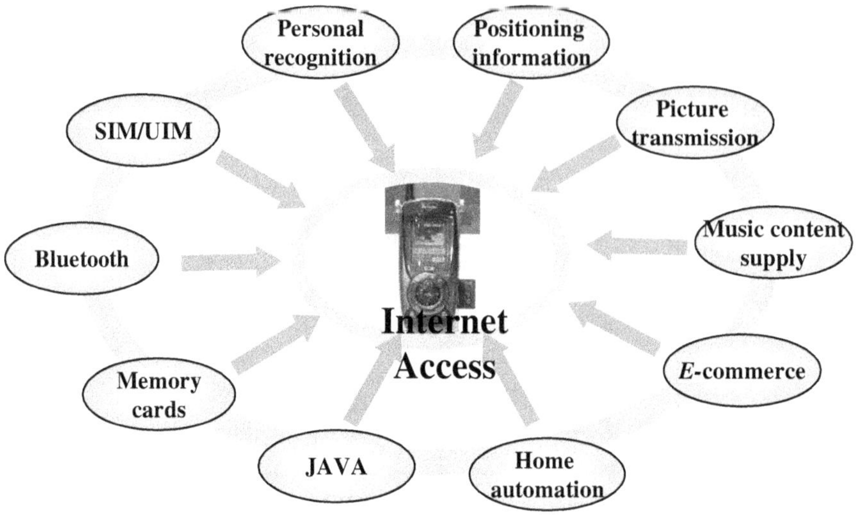

(a)

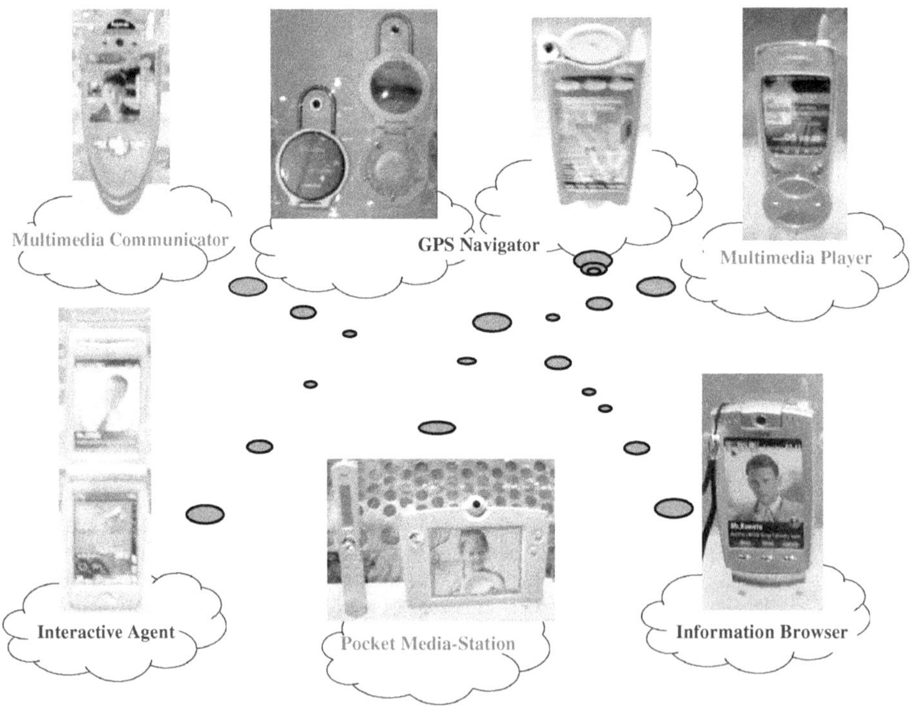

(b)

FIGURE 4.8 (*a*) New services drive proliferation of handset features and form factors. (*b*) New form factors and wireless service platforms.

- Standard Phone

Board Top Surface Area:	35 sq cm, mostly single-sided assembly
Component Pitch:	Mostly 0.8 mm CSP
Average Density:	20 pads/sq cm
Peak "Per Sq In" Density:	60 pads/sq cm
Peak "Per Comp" Density:	110 pads/sq cm

- Watch Phone

Board Top Surface Area:	12 sq cm, double-sided assembly
Component Pitch:	Mostly 0.5 mm CSP
Average Density:	40 pads/sq cm
Peak "Per Sq In" Density:	80 pads/sq cm
Peak "Per Comp" Density:	310 pads/sq cm

- PDA Phone

Board Top Surface Area:	55 sq cm, mostly single-sided assembly
Component Pitch:	0.5 mm CSP for low I/O, 0.8 mm CSP for high I/O
Average Density:	35 pads/sq cm
Peak "Per Sq In" Density:	80 pads/sq cm
Peak "Per Comp" Density:	220 pads/sq cm

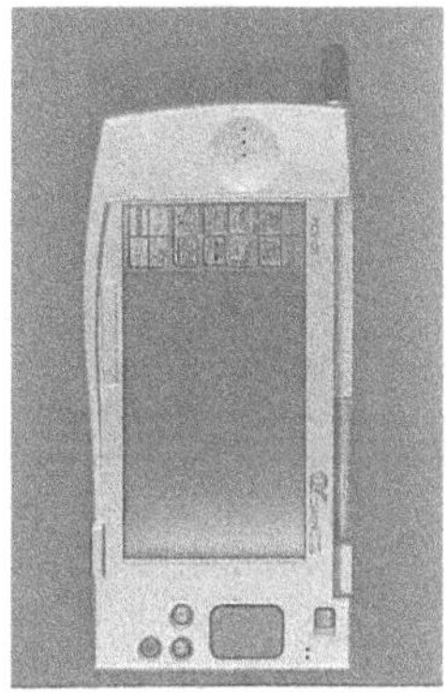

Source: Prismark Partners

FIGURE 4.9 Mobile phones in 2004.

as quickly as possible, thereby leaving standards in the dust.[9] Adopting a universal specification for an internal component may not be as critical as it is for end products such as VCRs, but a uniform Flash plus SRAM S-CSP specification surely would serve to facilitate expanding the industry volume readiness and infrastructure cost reductions. The same will hold true for the emerging integration of ASIC plus memory S-CSP packages. Here, most digital signal processing (DSP) and baseband processor applications can look to existing FBGA standards to adopt package outlines and pin-outs that are supported by open tool package, tray, and test socket offerings. This wide range of square FBGA packages can accommodate the logic plus memory stack size requirements without the need to invest in a new set of standards or infrastructure that was required for stacking of Flash plus SRAM in the initial S-CSP offerings. The packaging suppliers will need to develop the wafer thinning, thin die attach stacking, and low loop wire bonding and design rules to accommodate two dies in a package originally designed for one.

2. Widespread availability of thin, high-density PWBs using microvia fabrication and thin-core technologies is essential for both the needed buildup multilayer handset product boards and the thin two- to four-layer interposer substrates in laminate based S-CSPs. Substrates using 100-μm and thinner BT cores with 50-μm trace and space, 130-μm bond fingers, laser via in-pad, and tight soldermask registration control are critical to meet the advancing Si integration, density, I/O, and thin package height requirements emerging for new high-density memory stacks and ASIC plus memory stacks.

Note: Prismark forecasts the demand for microvia and HDI substrates used in product boards and packaging substrates to increase from $2.1 billion in 1998 to $7.6 billion in 2002, with 67 percent of this total consumed by the packaging substrate applications.[10] Substrate suppliers should plan aggressively to increase their capacity and capabilities to capture share in this high-growth market.

3. Widespread availability of cost-effective high-density single- and two-metal polyimide flex or TAB circuits for tape-based S-CSPs. The TAB and flex circuit industries need to invest aggressively in new one- to two-metal capabilities and capacity or will risk losing share to the emerging rigid thin-core suppliers.

4. A value-chain appreciation for the packaging and interconnection industries. Total cost-of-ownership valuations are required to support the higher levels of R&D and capital equipment required to meet the current wave of advanced 3D and packaging integration requirements. Thus 3D packages such as S-CSP, measured on a cost per pin basis, fail to account for the total cost of ownership benefits S-CSP delivers to the OEM.

5. Greater rate of attraction for students to pursue engineering and technical degrees in electronics packaging and interconnection disciplines.

These issues must be resolved in a cost-effective and timely manner to support the huge demand for S-CSP solutions required to support the current and projected handset, PDA, and memory block applications. Forecasts usually vary (especially for emerging technologies such as S-CSP), but current estimates require that the industry plan on capability and capacity to support up to 300 million S-CSP units for 2001. Equipment, test, material, and service suppliers cannot afford to wait until the demand/forecast outlook is clearer before finalizing their S-CSP strategies. Thus Sharp, Amkor, and others have developed flexible, robust S-CSP technology platforms and roadmaps to support a diverse range of new die combinations and end-application requirements. Figure 4.10 illustrates technology capabilities or building blocks Amkor has developed to provide a foundation to support the development and industrialization of 3D, SiP, and application-specific packaging platforms.

4.6 THREE-CHIP INTEGRATION IN S-CSP PLATFORMS

To understand the benefits that triple-chip stacking can provide to handsets requires a quick review of Figs. 4.3 and 4.4. Figure 4.3 shows a flattening in the rate of weight reduction below the 100-g threshold broken in 1996. As can be deduced from Fig. 4.4, further weight, size, and cost reductions for next-generation handsets can be accomplished through stacking the Flash plus SRAM combination on top of the DSP function in ASIC 1. This three-chip stacked package and wire-bond structure is represented in Fig. 4.11*a* and *b*.

For three-chip memory stacks (where die sizes and bond pads have been optimized for stacking), by using a thin, flex tape–based interposer and thin backgrind wafer technology, the triple-chip S-CSP can still meet the 1.4-mm maximum thickness goal associated with advanced handsets.

The adoption of triple-chip S-CSP is being pursued by Japanese handset makers to allow them to cost-effectively crack the 60-cc and 60-g benchmark as represented in Fig. 4.12. In addition, three-chip

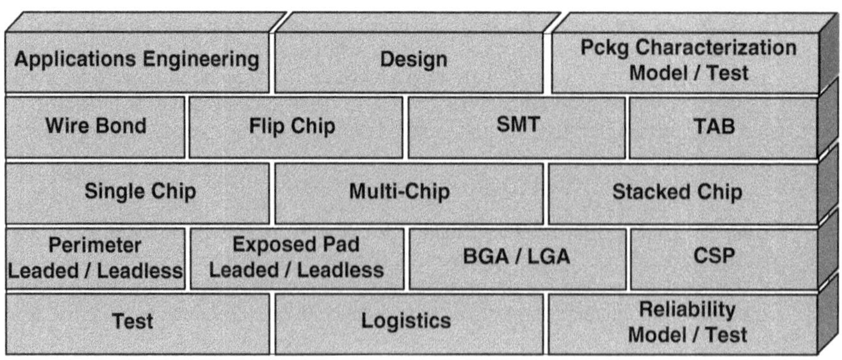

FIGURE 4.10 Capabilities required for an SiP foundation.

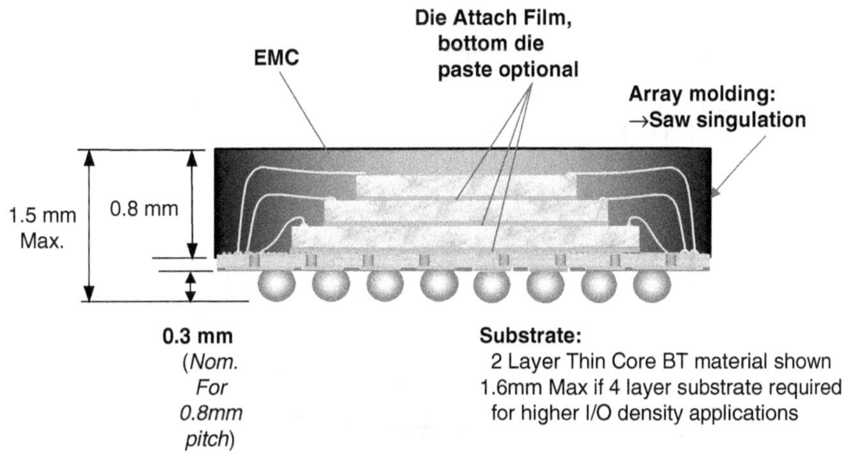

(a)

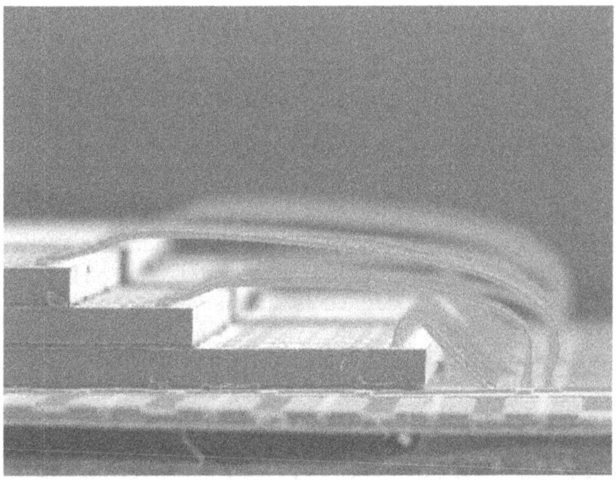

(b)

FIGURE 4.11 (*a*) Die-stacked CSP. (*b*) Die-stack wire-bond photo.

stacking will help enable high-end/data-centric cellphones such as the I-Mode and 3G handsets to be delivered with more features and functions in the small form factors consumers demand.

Another multichip 3D package that is gaining strong adoption for three-chip integration requirements is the so-called stacked MCM (S-MCM) depicted in Fig. 4.13. Here, die sizes, I/O, and wire-bond complexity do not allow integration in a three-die pyramid stack, so one of the dies is mounted to the side of the stack to accommodate thermal enhancement or fine-pitch wire-bonding requirements. Table 4.1 provides a comparison of the benefits and design flexibility for three-chip pyramid versus the 2 + 1 configuration associated with the stacked MCM. Figure 4.14 illustrates the various die stack and placement configurations feasible with stacked multichip packaging.

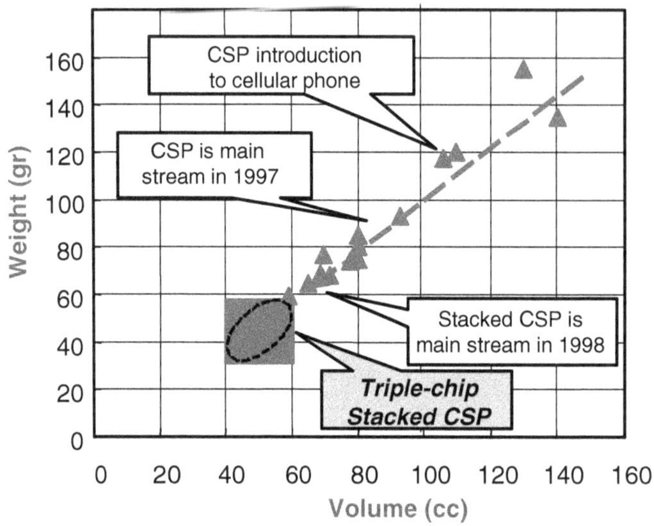

FIGURE 4.12 Triple-chip S-CSP contribution to compact, lightweight cellular phones.

4.7 FLIP CHIP (FC) AND S-CSP TECHNOLOGY ROADMAPS

A great deal of recent attention, particularly for analog and memory devices, has been placed on FC direct chip attach (DCA) or wafer-level packaging technologies for improved Si efficiency. However, as Fig. 4.15 illustrates, the best level wafer-scale or bare-chip solutions can achieve is 100 percent silicon efficiency, defined as the silicon area being equal to its associated package area. When you add the extra test and processing costs and underfill bleed buffer and rework limitations associated with current FC and wafer-level packages, they tend to be much less effective and harder to adopt than the 80 percent efficiency associated with leading single-chip CSPs. Further, the use of triple-chip stacking in emerging S-CSP technologies allows a level of 200 percent Si efficiency to be approached without having to tackle the more complex second-level interconnect and solder-joint reliability challenges associated with high-I/O 0.5-mm and finer ball pitches.

When this Si efficiency factor is looked at in combination with the widening gap between IC fabrication and PWB density capabilities, it becomes evident that 3D package integration is critical in ensuring that system suppliers can continue to leverage the unmatched rate of technology doubling represented by Moore's law. The International Technology Roadmap for Semiconductors (formerly SIA roadmap) identifies packaging as a critical enabling technology and calls for further investment and innovation in IC packaging so that packaging can scale with Si. One of the key functions of an IC package is to provide a space/density transformer between the submicron scale of the wafer-fabrication technologies and the submillimeter scale of the PWB technologies, but from a 2D perspective these technologies are not even on the same scale. One of the key enabling technologies to allow packaging to scale with Si without outpacing the PWB and printed wiring assembly capabilities to handle it is through use of the *z* dimension through stacked 3D packages that provide a volumetric density solution.

Figure 4.16 depicts the range of 3D packaging technologies being developed or possible within Amkor's diverse laminate-based package family to deliver higher levels of silicon efficiency to a wide range of applications. In addition to this 3D applications roadmap, Amkor has a range of stacked-die configurations available in standard and advanced lead-frame packages.

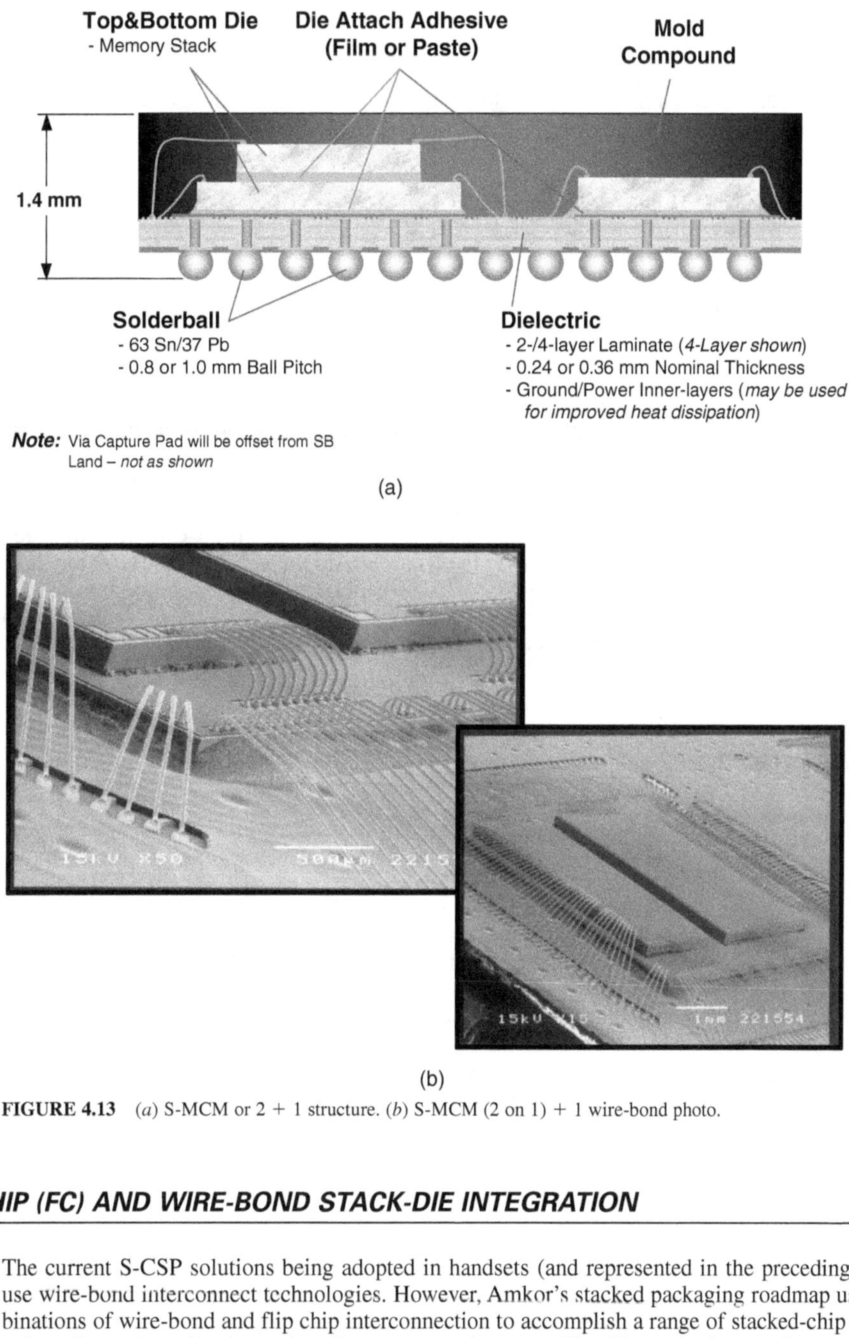

FIGURE 4.13 (*a*) S-MCM or 2 + 1 structure. (*b*) S-MCM (2 on 1) + 1 wire-bond photo.

4.8 FLIP CHIP (FC) AND WIRE-BOND STACK-DIE INTEGRATION

The current S-CSP solutions being adopted in handsets (and represented in the preceding figures) use wire-bond interconnect technologies. However, Amkor's stacked packaging roadmap uses combinations of wire-bond and flip chip interconnection to accomplish a range of stacked-chip configurations for various die size and performance requirements. The flip-stack, face-on-face structure is being developed to support rf applications through low-inductance/capacitance FC interconnections between rf and ASIC devices. This configuration can enable integration of GPS (global positioning system) receive/transmit radios in a small, low-cost form factor that can be integrated into cellphones

TABLE 4.1 S-MCM (2 + 1) versus Three-Chip Pyramid

Pros 2 + 1	Cons 2 + 1
Shared material set with two-die S-CSP, better substrate availability	Larger body size may be required, increasing board mount area and weight
Die shrink flexibility	
Use of 7-mil and thicker die	
Flexibility for different bond layouts, ease of routing, and die-to-die bonding	
Better thermal management (>1 W)	
Supports higher density requirements, two to four metal layers; easier to integrate passives	
1.4- to 1.2-mm height compatible	
Package, test socket, and tray infrastructure available	

and PDAs. This package structure also can be used to support Bluetooth radio components for a wide range of short-range wireless networking applications. The FC plus wire-bond device stack can be an effective solution for ASICs and graphics processors that are looking to integrate higher memory capacities than embedded economics can support. See Fig. 4.17 for cross sections of FC plus wire-bond die stacking. These interconnect advancements and the S-CSP infrastructure being established to support initial applications in Flash plus SRAM through ASIC plus memory for cellphones can be applied to a wider range of device combinations, package platforms, and end-market applications. Die stacking along with emerging thin package stacking such as Amkor's *et*CSP and Toshiba's paper-thin packaging will allow 3D packaging to deliver continuous improvements in optimal cost and performance for the new millennium.

4.9 CONCLUSION

Sharp Corporation and Amkor Technology have aligned S-CSP technologies to ensure that current Flash plus SRAM high-volume requirements are addressed from a global perspective. Further, the two industry leaders have reviewed and aligned their packaging roadmaps to ensure that the size, weight, and cost-reduction benefits of stacked-chip packaging technologies are extended to new device combinations and applications. Stacked-chip interconnection and BGA and CSP technology platforms are merging to create a forceful fourth wave of packaging, characterized as 3D package integration, that is critical in helping sustain Moore's law. The huge volumes and high rates of 3D packaging adoption from the mobile phone market are driving further enhancements and cost reductions in stacked-packaging technologies. Figure 4.16 illustrates both die and package stacking advancements that are being developed and adopted by a growing range of portable or memory-intensive applications. 3D packaging is becoming a critical strategy for many OEMs and integrated device manufacturers (IDMs) as a silicon and system-in-a-package integration tool as well as a cost-effective approach for stacking versus embedding large memory blocks.

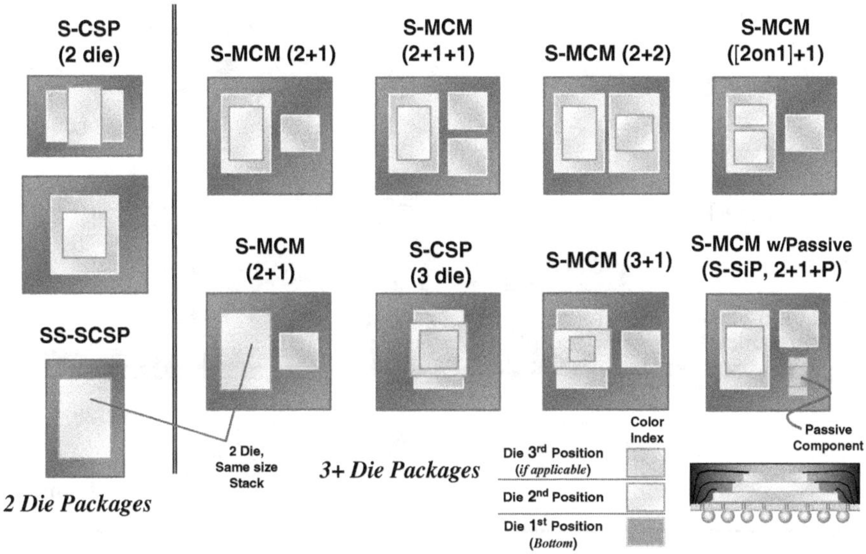

FIGURE 4.14 Stacked-package–die-position designations.

Source: Sharp

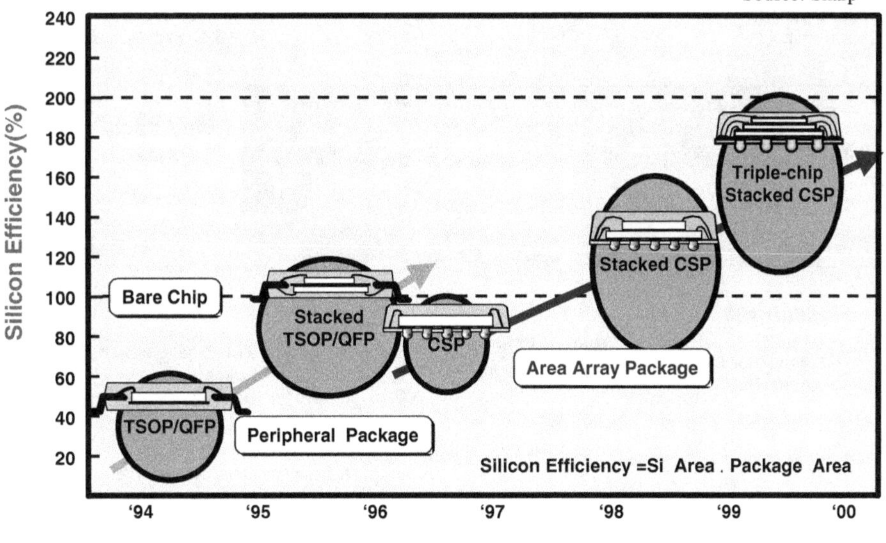

FIGURE 4.15 Silicon efficiency.

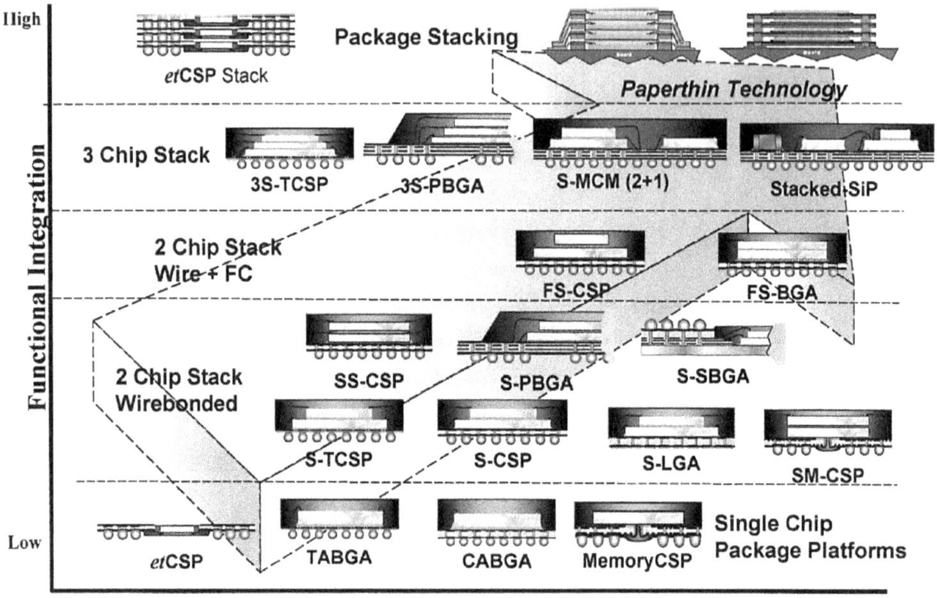

FIGURE 4.16 3D packaging application roadmap.

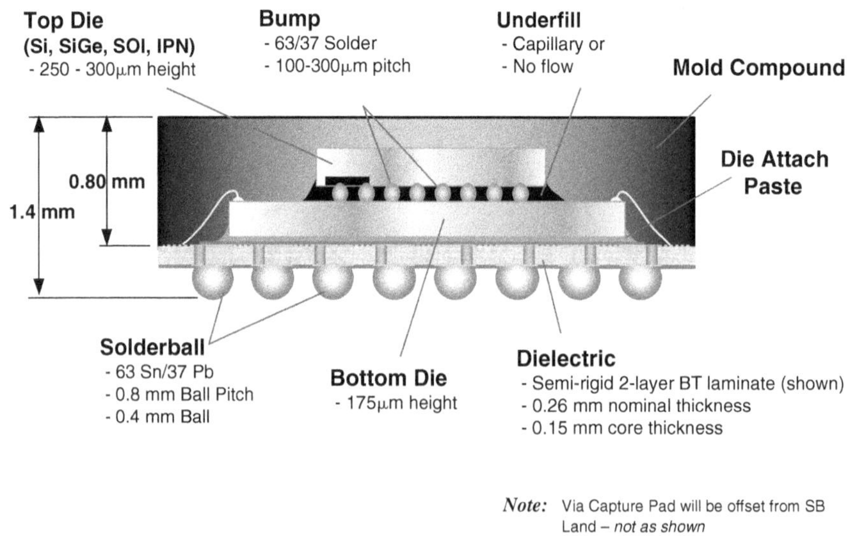

FIGURE 4.17 Flip-stacked CSP (fsCSP).

4.10 REFERENCES

1. B. Swiggett, Amkor Technology's Third Annual Customer Symposium, Santa Clara Marriott, Santa Clara, CA, November 1999.
2. B. Prior, "Trends in Electronics Packaging," SMTA Silicon Valley Chapter Meeting, August 1999.
3. J. Grenier, "The State of the Semiconductor Industry," IPC AMRC Conference, December 1998.
4. J. Woodyard, L. Smith, and K. Ramakrishna, "Stacked-CSP Advances in Handheld Electronics," *European Semiconductor Magazine,* Back-end Supplement, November 1999.
5. M. Kada, "Stacked CSP: A Solution for System LSI," *Chip Scale International,* 1999, pp. B-1–B-7.
6. TechSearch International, Inc., BGA/CSP Development Update Service, 2000.
7. E. Cuellar, "MultiMediaCard (MMC): World's Smallest Removable Storage Device," in *Proceedings of the Wireless Symposium Portable by Design Conference,* Boston, MA, Penton Media, Inc., September 1998.
8. H. Ueda, Amkor Technology Japan, Amkor internal report on mobile phone and associated packaging trends driven by I-Mode handset/service adoption in Japan.
9. B. Bonner, 1999 Flash Memory Applications Report, Dataquest Market Trends Report, Gartner Group, April 3, 2000, p. 28.
10. Sections of this chapter are taken or expanded from L. Smith and M. Kada, "Advancements in Stacked Chip Scale Packaging (S-CSP) Provides System-in-a-Package Functionality for Wireless and Handheld Applications," SMTA Pan Pacific Microelectronics Symposium, January 2000.

CHAPTER 5
COMPLIANT IC PACKAGING

Joseph Fjelstad and Gary Yasumura
Pacific Consultants, LLC.

Young Gon Kim
Tessera, Inc.

5.0 INTRODUCTION

Packaging silicon integrated circuit (IC) chips at or near the size of the chip presents very important electromechanical challenges to the electronics system assembler. The mismatch in coefficient of thermal expansion (CTE) creates significant stress on the interconnections made between the packaged device and the substrate to which it is mounted. Historically, making the leads of the packaged device flexible or compliant has served this need.[1]

Making the IC package leads compliant serves to compensate for this mismatch, making the package capable of meeting the demands of the full range of environments that the electronic assembly may encounter in use. If the package lacks a stress-management mechanism, the package assembler/silicon provider must limit the warranty of the package to only certain environmental conditions.[2]

This chapter will review a packaging concept that provides just such compliance as is necessary, obviating the need for an expensive underfill process to protect the solder joints of the chip-scale package (CSP). The packaging concept that is the subject of this chapter is one developed by Tessera, Inc. It has become one of the most recognizable and popular formats for CSPs in the world. The μBGA (microBGA) CSP is presently being used extensively in Flash memory applications, having been adopted by Intel and others for this purpose.[3] It is also the reference package for Rambus RDRAM, which is expected to dominate in next generation computer memory applications due to its fast operating speeds, which are projected to 1.6 GHz.

The fundamental structural elements of the Tessera μBGA CSP are shown in Fig. 5.1. Some licensees of the technology (e.g., Amkor, Hitachi, Sony, and others) have made modifications to the manufacturing process, but the fundamental elements remain the same.[4] There are a number of special attributes to the μBGA CSP. These attributes help to improve the performance of electronic devices packaged in μBGA format, but the provision of a strain-reduction layer of material between the flex tape and the silicon die is perhaps the most important.[5] This patented[6] technique affords a CTE mismatch strain buffer between the die and the substrate to which the device is to be mounted. It will be reviewed in more detail later in this chapter.

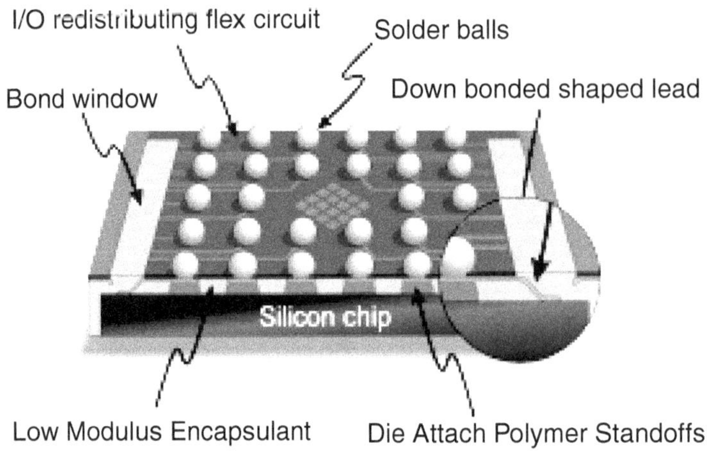

FIGURE 5.1 Construction elements of the basic μBGA CSP.

5.1 PACKAGE TECHNOLOGY OBJECTIVES

The μBGA CSP was developed in response to two different conditions that were in effect in the early 1990s. One was the lack of a readily available good die, a problem that has been a continuing concern for the multichip module (MCM) industry. The other was the fact that bumped dies for flip chip applications were not readily available. A related concern was the fact that bare dies were not standardized. This meant that each die had an individual footprint, making the prospect of standardized component libraries for printed wiring board (PWB) design impossible. These conditions combined to drive the concept of packaging ICs at chip size.

The μBGA package was developed originally for a unique MCM in development at Tessera. While electrical and thermal performance were key concerns,[7] compliance was considered to be very important as well. In fact, the μBGA CSP was originally called the *Tessera-compliant chip* (TCC). Compliance was expected to provide some relief to the electrical test process by making the device contact vertically deformable, thus facilitating reliable contact with test sockets. Testing would ensure that the dies were good at the time of assembly. In short, the package was designed to bring together the benefits of flip chip with those of standard surface-mount technology (SMT) packages (see Fig. 5.2).

5.2 I/O PLACEMENT

Array placement of input-output (I/O) terminals is the most economical use of space and is the method most favored for CSPs. Area array packaging concepts are intrinsically powerful. They allow I/Os to be positioned at a relatively course pitch, and yet they consume much less area than peripherally leaded packaging alternatives with finer lead pitches (see Fig. 5.3). Still, very small dies with low I/Os may be better served by placing I/Os around the edge of the die.

5.3 COMPLIANT CSP CONSTRUCTION

Tessera's compliant chip packaging technology went through a number of evolutionary changes from the original concept to reach its present state of development. Following is a review of the innovation path to the present product.[3]

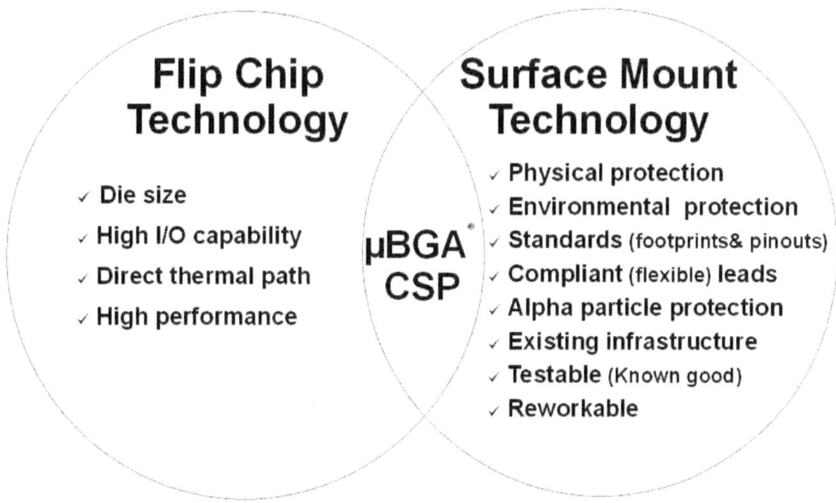

FIGURE 5.2 A simple graphic of the conceptual μBGA.

FIGURE 5.3 μBGA CSP compared with the TSOP and SOP packages. Note that the lead pitch on the μBGA is greater than that on the TSOP. (*Courtesy of Intel.*)

5.3.1 Materials of Construction

Very few materials are required to manufacture the μBGA CSP. However, each of these materials plays a significant and vital role. Primary elements include

- *Flexible base-film material.* A nonreinforced flexible base film is used as the foundation in the construction of the μBGA. The primary reason for using flexible films is that they have a lower Young's modulus than the reinforced and rigid alternatives. The low modulus contributes to strain reduction on the interconnections made at the next level (normally a solder ball).
- *Circuit redistribution wiring.* The metal circuit traces provide for the redistribution of chip I/Os from the bond pad locations to the locations required for component mounting and interconnection

at the next level. The base metal is normally copper, but gold and aluminum leads also have been employed in such constructions.

- *Buffer or compliant layer.* A critical element of any compliant CSP construction is the compliant layer, also called a *buffer layer.* This is the layer that absorbs most of the strain caused by CTE mismatch during thermal excursions related to either operation of the electronic product itself or the environment in which it is operated. The result is that the interconnection to the next level (such as a solder ball) is protected. This layer theoretically can be made to vary in thickness directly with its elastic modulus, but because of system complexity, it is normally formulated to provide strain relief at a thickness of approximately 75 to 150 μm. The compliant layer also serves to protect the metal interconnection leads or flexible links. A very important requirement of the buffer layer is that it be free of alpha-particle emitters. This is so because if the buffer layer is to be used directly on the active circuit side of the die, it is a necessary condition that it be free of alpha-particle emitters. (This has been a concern in memory for many years, but as feature sizes are reduced, it becomes even more critical.) Silicone elastomers have proven highly effective in this role; moreover, because moisture moves rapidly through the material, the package can be made moisture-insensitive.
- *Flexible link.* The flexible link serves to make reliable electrical and mechanical interconnection between the chip and the flexible CSP interposer. In the original concept of the μBGA CSP, this link was to have been a gold bond wire. However, a follow-on concept that made the flexible link an integral element of the redistribution wring obviated the need and made the package higher performing, smaller, and lower cost. The lead must be strain relieved by the compliant encapsulant/buffer layer as well to protect it from early failure due to local work hardening.
- *Next-level contacts.* A wide range of options are available for contacting the μBGA CSP; planar lands, electroformed bumps, and solder balls have all been examined. Bumps or balls can facilitate testing and allow for easier interconnection to the next-level interconnection structure (e.g., PWB).

Obviously, many variations can be envisioned as to the specifics of the materials used in such constructions (Fig. 5.4).

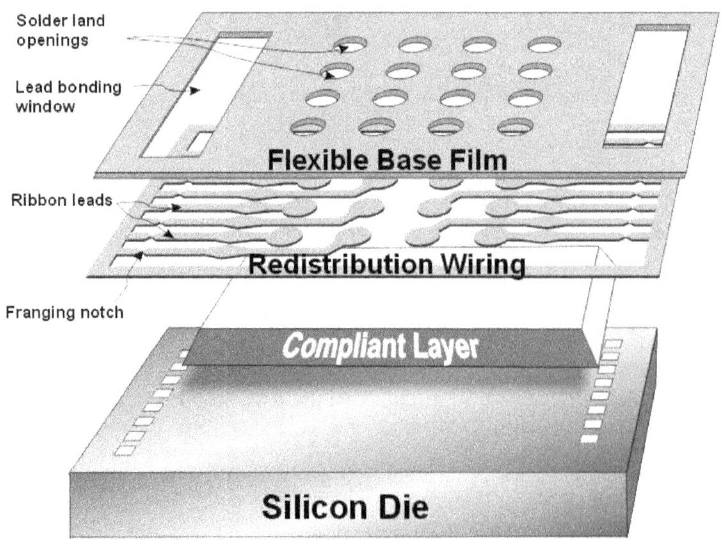

FIGURE 5.4 Elements of construction of the compliant μBGA CSP.

5.3.2 Bond Lead Design

Lead design is of great importance to successful design of the ultimate reliability of the package.[8] The bond lead must be able to move in all three dimensions of space. Lead design also must accommodate the needs of manufacturing in order to facilitate lead bonding and shaping, which have a direct impact on design. Lead design is also heavily influenced by the other elements of construction, such as whether the lead is placed on top or bottom of the flexible film. Finally, the lead design is also affected by the thickness and physical properties of the encapsulant. The key elements of lead design in three dimensions are provided in Fig. 5.5.

5.3.3 Principles of Operation

The physical principles of operation of the μBGA-compliant CSP product group are relatively simple. A compliant layer is provided between the die and the flex substrate to which the die package is mounted. The device has been engineered to protect the most vulnerable element of a mounted IC package—the solder joint. The provision of a compliant layer within the package in combination with the flexible link or bond lead serves this purpose. Figure 5.6 illustrates the concept.

This solves the problem of CTE mismatch between silicon chip and PWB by using a compliant encapsulant between die face and interposer. This takes the stress off the solder balls, which normally fail first on most CSPs. Illustrated in Fig. 5.6 is the die face-down implementation of the compliant-layer concept.

5.3.4 Other Compliant μBGA Configurations

The die face-down construction of the compliant μBGA CSP is widely known and in use around the world, but it was not the original design. Wire-bonded compliant packages preceded it in conception. Following are brief descriptions of earlier Tessera-compliant chip constructions (Fig. 5.7):

- *Die face-up.* Dice face-up is by far the most common method of presentation of the die both to the package for assembly and to the substrate on which the assembled IC package is mounted. It was

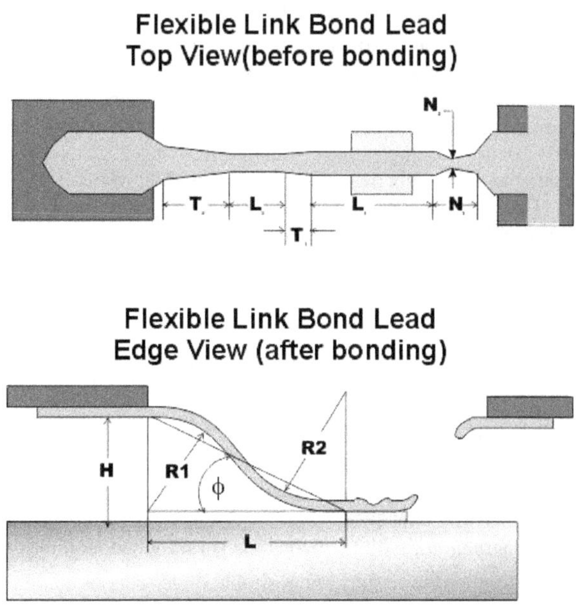

FIGURE 5.5 Design elements of the flexible link/bonding lead for the μBGA CSP.

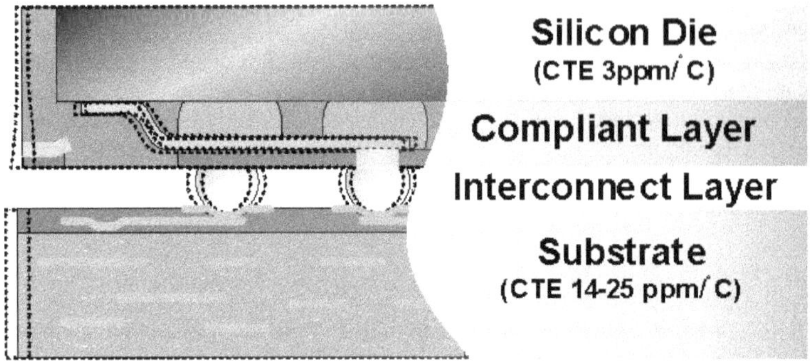

FIGURE 5.6 Structure of the μBGA CSP.

this concept that was first considered for the μBGA CSP. Two variations were examined. One was a flat, flexible film (trade named FμBGA) that used a much thicker than normal die-attach material of a modulus sufficiently low enough to reduce the stain on the solder ball beneath. Most flex circuit–based CSPs such as TI's MicroStar BGA originally used a traditional thin die-attach layer and reported lower reliability. However, TI reported in the *Proceedings of the 1999 Pan Pacific Electronics Conference* that better results are seen when a compliant layer is added.

- *Wrap-around.* This concept was examined in order to avoid having to make small holes in the flex tape. The die was mounted on the side opposite the circuits of the flex circuit, again using a thick, low-modulus die attach. The flex was then wrapped around a support ring, and the I/O bond pads of the die were wire bonded to the flex circuit. Other investigators, including Fujitsu, have reported exploring this concept since it was first examined at Tessera. However, the structure is difficult to assemble and is not a likely candidate for mass production.
- *Die face-down.* Another wire-bonded variant of the compliant μBGA CSP was described in an early patent. The original structure was the inspiration for today's version of the μBGA package. Some contract manufacturers have adopted the method as a means of performing wafer-level assembly as described in early Tessera patents. Toshiba is using the method for discrete chips to package memory. The structure resembles the standard μBGA package except that a wire bond is used in place of the integral lead. This design used wire bonding to make a connection from the interposer to the chip. The patented face-down wire-bonded structure is being used by some contact IC assemblers at present.

5.3.5 I/O Configurations

There are several different possible configurations of the compliant μBGA package. Following is a review of the most commonly used constructions. The package types are illustrated in Fig. 5.8.

- *Fan-in.* All I/Os are placed within the die area and interconnected by redistributed wiring to die bond pads located at the perimeter of the die.
- *Center bond pad.* Like fan-in, all I/Os are placed within the die area and interconnected by redistributed wiring, but die bond pads are located in the center of the die, a placement favored for memory devices. This construction easily facilitates die shrink when required.
- *Fan-in/fan-out.* Fan-in/fan-out constructions are desirable under certain conditions, such as when a peripheral lead die is shrunk to a point that makes it impossible to hold the solder ball I/O footprint on the face of the die or when a larger I/O pitch is desired.

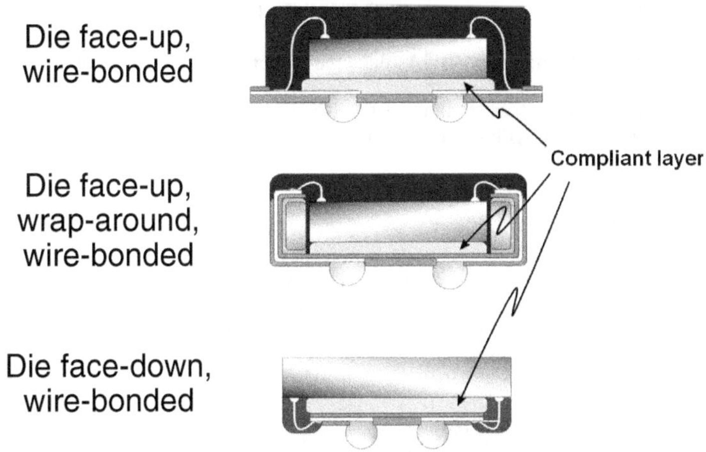

FIGURE 5.7 Early concepts of the compliant CSP.

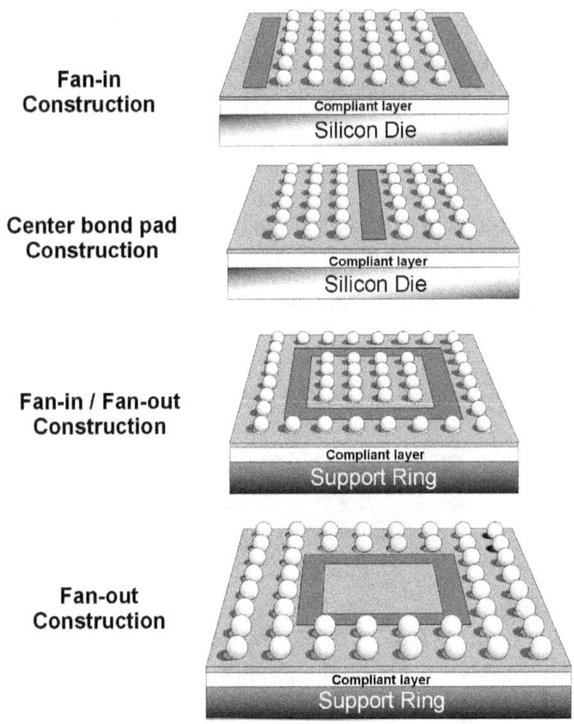

FIGURE 5.8 The compliant layer.

- *Fan-out.* Fan-out-only constructions of compliant packages are used when the die is very small and yet has a significant number of I/Os for its size. Fan-out design also can be applied to the replacement of rigid ball and array (BGA) packages when better mechanical compliance is desired.

The layer is a key characteristic of the μBGA package and is incorporated into all the various construction formats possible, from chip size to larger than the chip, to meet the needs of the industry infrastructure.

5.3.6 Multiple Metal Layers

The vast majority of CSP applications at present only require one metal layer for I/O redistribution, but with expected increases in complexity, two, three, or more metal layers may be required to meet next generation performance demands. Some products such as SRAM and next generation fast DRAM such as RDRAM are looking forward to such need. An example of a two-metal-layer fast SRAM device is shown in Fig. 5.9.

5.3.7 Adapting to Die Shrink

Die shrink is standard practice in semiconductor manufacture. It represents a challenge to all CSPs. CSP size obviously is determined by the die size. Thus, if an area array format is chosen for I/O presentation to the next-level assembly, die shrink translates to package outer dimension shrink as well. If the I/Os remain within the perimeter of the die at the chosen pitch, no difficulty is expected. However, if die shrink results in I/Os being located outside the perimeter of the die, a decision must be made to either shrink the I/O pitch or make the package larger than otherwise would be necessary.

Thus, for CSPs, it is necessary to establish a design strategy for the package outer dimension and to determine whether the old package outer dimension will be used or a new reduced package outer dimension will be established. Die shrink is especially challenging to die face-down constructions that use wire-bonding technologies. This is so because wire-bond pad placement often interferes with next-level I/O contact land location. Anticipating all future die shrinks and using them for establishing I/O placement solves the problem, but the problem is then transferred to the next-level PWB design, manufacture, and assembly, which can add significantly to the manufacturing cost of the assembly. Despite this problem, clever design layout and process modification can nearly subdue die shrink challenges. It is possible to stave off some of the die shrink problem by adopting a fine-pitch ball matrix from the beginning. Figure 5.10 illustrates methods for addressing die shrink for face-down constructions.

FIGURE 5.9 A larger than die size fan-in/fan-out μBGA chip package.

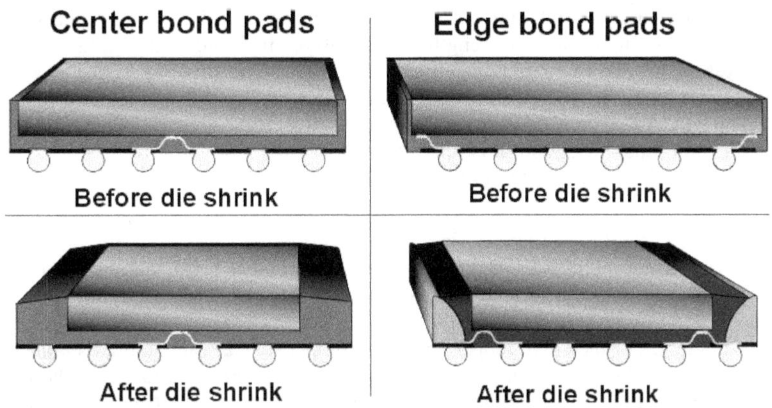

FIGURE 5.10 μBGA CSP die shrink strategy varies with location of the bond pads.

5.4 MANUFACTURING PROCESSES

A number of different manufacturing processes have been developed over the years to build the μBGA CSP. Each new process resulted in significant improvement in terms of both yield and cost reduction. While the prime objectives for the technology have remained unchanged (i.e., highest performance in the most reliable package possible), the latest process technology also excels in the area of cost reduction. The manufacturing process for the μBGA is capable of producing IC packages at less than $0.0049 per lead according to cost modeling data. As a point of reference, most IC packages are presently priced at approximately $0.01 per lead.

Even though earlier manufacturing methods did not meet the cost target, system developers found that their overall cost-performance needs nevertheless could be met by the μBGA CSP in many cases. This is so because the package could be placed at rates of up to 20,000 units per hour in assembly at very high yield, and it did not require an expensive underfill process after it was mounted on the PWB to protect the delicate solder joints.

5.4.1 Original Process

Once settled on the idea of an integral lead frame with bond ribbon ultimately being superior to wire bonding, an attempt to find a suitable process was undertaken. The fundamental notion was to have tape automated bonding (TAB)-like lead frame that fanned the fine-pitch I/O of the die bond pads inward to a relatively coarse area array of contacts rather than outward to a fine-pitch peripherally leaded device. Many challenges were in store. For example, one of the long-standing problems with TAB was the fact that the delicate cantilevered inner leads could be bent or damaged easily. If these same leads were to be cantilevered around the outer edges of the device, the problem likely would be worse. Because of this concern, it was decided that the leads should remain attached to a carrier film until the moment of bonding. However, having the lead break loose for bonding reliably was a challenge. Originally, the leads were designed to be tenuously attached to the polymer carrier film and would be pulled away from the film during the lead bonding process. However, this method was not as consistent as was desired. The concept of a frangible lead based on a notch thus was developed.

Most of the world's tape suppliers shied away from the challenges presented by the tape design. As a result, it fell to the company to develop processes to manufacture the flex circuits, as well as for assembly. Early flexible circuit tape for such material was made at the company using electroformed gold leads, laser-drilled holes and windows, and electroformed nickel-gold bumps. Gold was chosen first because of its malleability and resistance to work hardening. These were deemed important attributes for structural reliability. Lead design matured over time, and a taper was added to the lead to create a "beam of uniform strength." This feature prevented lead kinking and made lead shaping during bonding (an important design element) easier. The structure of the original flexible circuit tape μBGA package is shown in Fig. 5.11.

The μBGA CSP construction has changed over time to improve manufacturing efficiency while maintaining reliability. Also illustrated in Fig. 5.11 are the major process revisions that have been employed in the manufacture of the μBGA CSP.

5.4.2 The Zinger Assembly Process

Once TAB tape suppliers decided to produce μBGA tape in small volume, the company was able to demonstrate a more cost-effective manufacturing method. The result was a process trade named Zinger. The new process relied, to the maximum extent possible, on the existing package assembly infrastructure. Off-the-shelf copper-based TAB tape, standard stenciling equipment, pick-and-place machines, wire bonders, dispense equipment, etc., all were used to create the manufacturing line. While the first process was not fully optimized, it did accomplish its objectives and enabled the first mass production of product (Flash memory chips) by technology licensees.

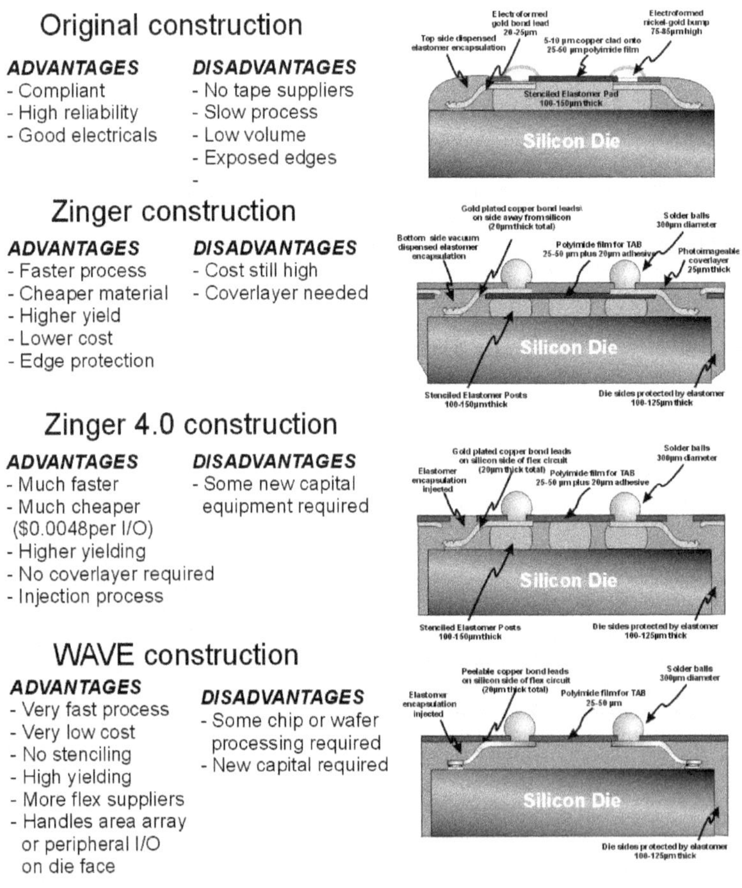

FIGURE 5.11 Construction changes over time.

- *Zinger 1.4 process.* The first process released for volume manufacture was the Zinger 1.4. This process used existing equipment with only minor modifications.
- *Zinger 3.0 process.* The second process released for μBGA CSP manufacture was the first to have equipment specifically designed for it. Innovations included in the process were the use of a novel vacuum dispenser of encapsulant.
- *Zinger 4.0 process.* The latest of the Zinger process family has further innovations and new, specifically developed manufacturing equipment such as is required for injection. This process brings the cost of the μBGA package down to \$0.0048 per I/O according to cost models, with further reductions possible. The process is illustrated in Fig. 5.12.

5.4.3 The WAVE Process

The wide area vertical expansion (WAVE) process was developed originally to serve the needs of designers who sought to lay out chip I/Os in an array fashion over the face of the die. The process

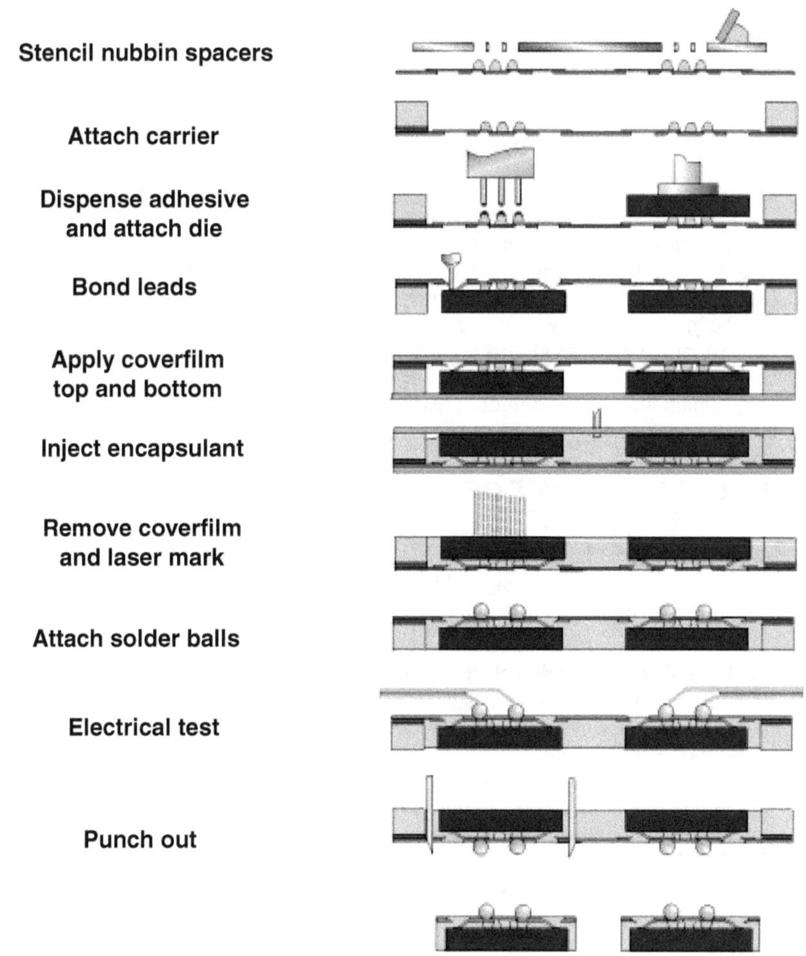

FIGURE 5.12 The Zinger 4.0 process.

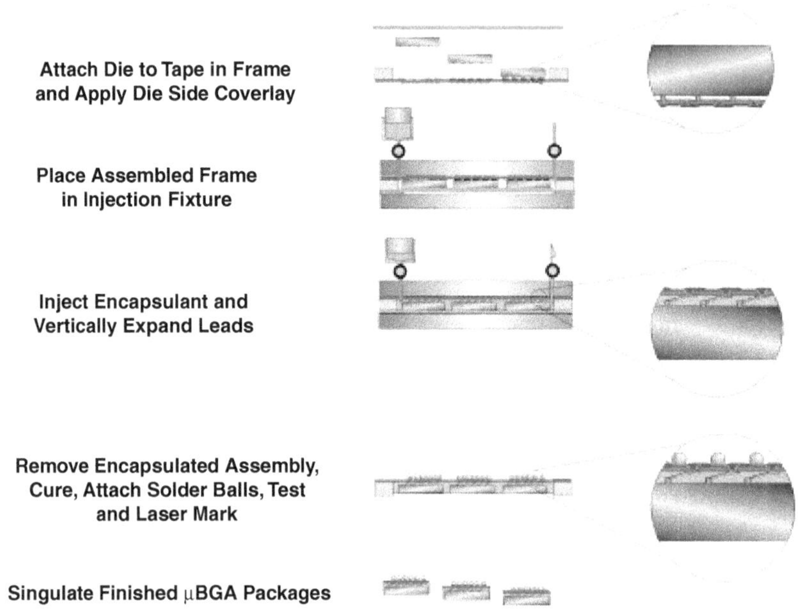

FIGURE 5.13 The WAVE process.

also was viewed as a likely solution to the underfill dilemma faced by flip chip technology users. It would in effect make flip chip devices into compliant packages. The process also was extended to include packaging directly on the wafer. The chip-at-time process is illustrated in Fig. 5.13.

The method greatly improves manufacturing efficiency by allowing for the injection of encapsulant. The method also greatly reduces the cost of chip packaging by allowing all interconnections to be made in a single step. The lifting of the leads is accomplished by injecting the encapsulant and curing it under pressure.[10] The process is suitable for both wafer-level and discrete IC packaging.

5.5 ELECTRICAL PERFORMANCE OF THE μBGA CSP

While the flexible circuit used in the construction of a μBGA CSP is capable of carrying high-speed signals from an IC chip through solder balls surface mounted to a PWB, a circuit designer can optimize the flex circuit's configuration and materials to ensure high-frequency signal integrity.

5.5.1 Software-Modeled *RCL* Parasitics

Modeling of the μBGA CSP has been performed by a number of different companies since introduction of the package to determine the common electrical parasitics of resistance, capacitance, and inductance. Pacific Consultants of Mountain View, California, used Ansoft's Q3D parameter extractor software to derive modeled *RCL* parasitics from the μBGA TV46 package. Four traces in the package were modeled. Solder balls at the beginning of each signal trace were assumed to be 0.37 mm in diameter. It also was assumed that these were surface mounted to an FR4 epoxy-glass board. The signal traces were 0.02 mm thick by 0.04 mm wide. The four bonded leads of the signal traces were bent downward and bonded to pads on the silicon die in accord with the manufacturing process. A length of 0.32 mm was used for each bond lead. While the standard TV46 does not have a ground plane, a future generation of the device is expected to incorporate one for performance reasons. Thus a ground plane situated above the polyimide layer was assumed for this model. The compliant elastomer layer fills the space between the signal traces and the silicon die. The four traces were modeled as "normal" conductors with a limited conductivity of 5.8×10^7 S/m and a dc analysis was carried out. The calculations include flux inside the conductors (Fig. 5.14 and Tables 5.1 through 5.3).[5]

Table 5.1[1] gives the thickness and dielecric constant values for TV46 while Table 5.2 indicates lengths of traces tested. Measured results are given in Tables 5.3, 5.4, and 5.5 that show resistance, capacitance, and inductance, respectively. Table 5.6 gives these values as predicated by Ansoft Field Solver.

The software's field solver automatically refines the mesh by optimizing the placement of new elements into the existing mesh. To gain the required accuracy for this model's inductance and resistance matrices, the solver created over 90,000 tetrahedral elements. The values are displayed in Tables 5.4 through 5.6.

The software-generated value of 3.1 nH results from calculating partial inductances composed of signal traces only. It is possible to use a "Reduce Matrix" command during postprocessing to include the ground planes in the loop inductances.

In a follow-on analysis, the conductors were modeled as "perfect" conductors, whose conductivity is infinite. The skin depth was an order of magnitude smaller than the cross-sectional dimensions of the signal traces. Perfect conductors are assumed in Ansoft because they provide a reasonably accurate solution. The measured values are displayed in Tables 5.7, 5.8, and 5.9 that give resistance, capacitance, and inductance for the ac case with ground. Predicated results are shown in Table 5.10 for comparison.

The ac case changes the *RCL* parasitics values. The resistance increases because current must travel in a smaller cross-sectional area and more heating occurs. Capacitance changes very little, as expected, and inductance drops by one third. Figure 5.15 visually displays these relationships.

5.6 THERMAL PERFORMANCE

With many CSPs, the backside of the IC chip is covered by plastic. However, with the μBGA CSP, the backside of the die can be accessed directly by a heat sink or thermal spreader. As a result, the μBGA package can provide excellent thermal performance. This is rapidly becoming a significant

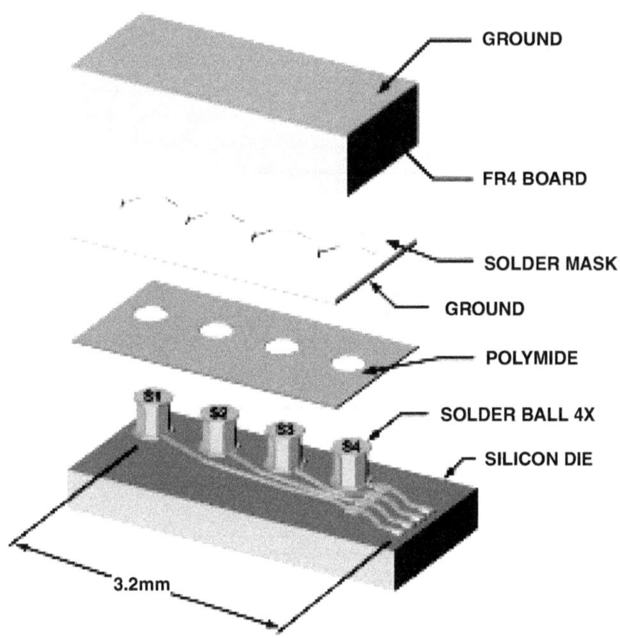

FIGURE 5.14 Test vehicle model.

TABLE 5.1 Layer Geometry

Layer	Thickness, mm	ε_r (relative dielectric constant)
Ground	0.02	NA
FR4	0.74	4.7
Solder mask	0.05	3.7
Ground	0.02	NA
Polyimide	0.02	3.3
Elastomer (not shown)	0.125	3.0
Silicon die	0.42	4.0

TABLE 5.2 Trace Lengths

Trace designation	Trace length, mm
S1	3.2
S2	2.5
S3	1.7
S4	1.0

TABLE 5.3 Resistance Matrix (ohms), Grounded Model, dc Case

	S1 trace	S2 trace	S3 trace	S4 trace
S1 trace	0.59	0	0	0
S2 trace	0	0.498	0	0
S3 trace	0	0	0.325	0
S4 trace	0	0	0	0.125

TABLE 5.4 Capacitance Matrix (picofarads), Grounded Model

	S1 trace	S2 trace	S3 trace	S4 trace
S1 trace	0.502	0.0389	0.00496	0.00176
S2 trace	0.0389	0.412	0.0328	0.00430
S3 trace	0.00496	0.0328e	0.308	0.0291
S4 trace	0.00176	0.0430	0.0291	0.185

TABLE 5.5 Inductance Matrix (henries), Grounded Model, dc Case

	S1 trace	S2 trace	S3 trace	S4 trace
S1 trace	3.10	1.41	0.728	02.77
S2 trace	1.41	2.55	0.921	03.20
S3 trace	0.728	0.921	1.63	0.379
S4 trace	0.277	0.320	0.379	0.723

TABLE 5.6 Ansoft Field Solver for dc Case for S1 Trace (3.2 mm length)

Ansoft field solver results	
S1 resistance	0.059 Ω
S1 self-capacitance	0.502 pF
S1 self-inductance	3.100 nH

TABLE 5.7 Resistance Matrix (ohms), Grounded Model, ac Case

	S1 trace	S2 trace	S3 trace	S4 trace
S1 trace	0.113	0.0658	0.00223	0.00247
S2 trace	0.0658	0.720	0.0234	0.000781
S3 trace	0.00223	0.0234	0.511	0.00955
S4 trace	0.00247	0.000781	0.00955	0.145

TABLE 5.8 Capacitance Matrix (picofarads), Grounded Model

	S1 trace	S2 trace	S3 trace	S4 trace
S1 trace	0.513	0.0396	0.00496	0.00174
S2 trace	0.0396	0.416	0.0331	0.00432
S3 trace	0.00496	0.0331	0.309	0.0293
S4 trace	0.00174	0.004.2	0.0293	0.188

TABLE 5.9 Inductance Matrix (nanohenries), Grounded Model, ac Case

	S1 trace	S2 trace	S3 trace	S4 trace
S1 trace	2.4	1.13	0.665	0.266
S2 trace	1.13	1.87	0.706	0.275
S3 trace	0.665	0.706	1.15	0.307
S4 trace	0.226	0.227	0.307	0.663

TABLE 5.10 Field Solver for ac Case for S1 Trace (3.2 mm length)

	Ansoft field solver results
S1 resistance	0.113 Ω
S1 self-capacitance	0.513 pF
S1 self-inductance	2.400 nH

concern as package wattage continues to climb due to increased operating frequencies, increases in integration, and reduction in feature sizes, causing them to become more resistive.

When applying the thermal spreader, a thin thermal pad is recommended. However, the pad should be capable of conforming well and should occupy all the space between the die and heat sink in a void-free manner.

θ_{jc} is thermal resistance of the system (package) and is given in °C rise per watt input. Low values are desirable since a minimum rise in temperature results in maximum component lifetime.

θ_{jc} can be as low as 0.8 and 0.2°C/W using a boron nitride thermal pad and zinc oxide thermal paste, respectively, as interfaces (Table 5.11). This configuration provides much better thermal performance than can be had with a cavity-down BGA.

The short thermal path has other benefits in that it facilitates both assembly and rework. Figure 5.16 illustrates this aspect of the package when compared with a plastic BGA. This illustrates how the package can facilitate both assembly and rework by requiring a lesser amount of energy to accomplish solder-joint reflow.

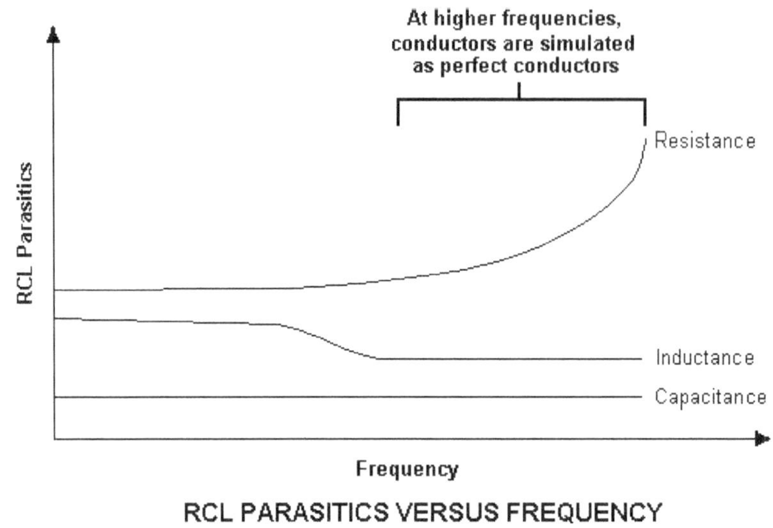

FIGURE 5.15 *RCL* parasitics versus frequency.

TABLE 5.11 Thermal Conductivity Measures of Various Heat-Sink-Enabling Materials

Material	Thermal conductivity (W/m·°C)	Pad thickness (mm)	θ_{ja} (°C/W)	θ_{jc} (°C/W)
Silicone rubber	0.4	0.20	5.8	3.8
Thermally conductive elastomer	2.2	0.20	4.2	2.2
BN-doped elastomer	6.0	0.15	2.8	0.8
Thermal grease	0.8	0.05	2.2	0.2

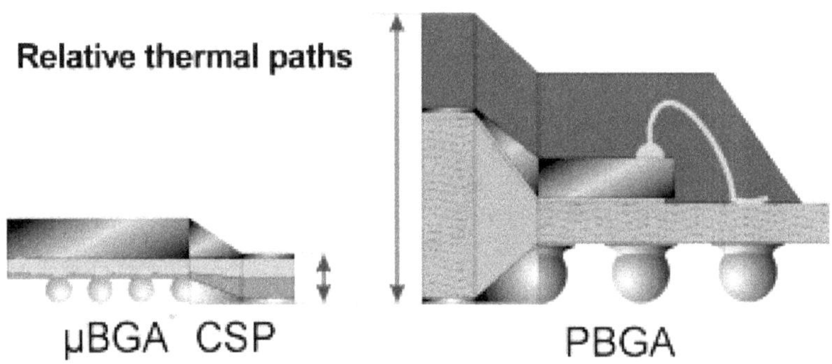

FIGURE 5.16 Relative thermal paths of the μBGA versus a plastic BGA.

5.7 RELIABILITY AND FAILURE ANALYSIS

The μBGA CSP has been proven to be a highly reliable IC package. It is considered to be among the most reliable packages ever produced. The μBGA CSP has demonstrated its reliability in numerous reliability studies by both OEMs and consortia.

Qualification testing to determine reliability normally is performed both on the package by itself to determine package-level reliability (PLR) and again with the device mounted on a substrate such as FR4 to determine board-level reliability (BLR). Qualification testing for reliability has long been an accepted practice to provide the end user with a high level of certainty that the product will not fail prematurely because of faulty package design or assembly processes.

Mil-STD-883 provides a detailed prescription for testing all types of electronic devices and has long served as the electronics industry's general qualification standard. While presently there is an effort underway to redefine qualification testing to a more benign regimen (e.g., 0 to 100°C for on-board reliability), the μBGA CSP was designed to withstand the rigors of the most stringent testing without paying a price penalty.

5.7.1 Moisture Sensitivity

Moisture can outgas explosively during reflow soldering, resulting in a cracked package. Because of this, moisture sensitivity of chip packages is an item of considerable concern to the electronics assembler. Silicon dominates the CSP, with encapsulants playing a much smaller role, and thus limits the total amount of moisture that can be absorbed. However, CSPs still can be affected adversely by moisture. The μBGA CSP, depending on the type of encapsulant used, is capable of meeting the requirements of JEDEC level 1. Level 1 is considered the most desirable designation because it implies moisture insensitivity. The JEDEC requirements are outlined in Table 5.12.

5.7.2 Conditions for Reliability Testing

The μBGA CSP has passed the full battery of tests outlined in Table 5.13. These conditions, while extreme, have been the benchmark for electronic packages for many years. The purpose of such testing is to stimulate failure so that device reliability can be determined and understood and reliability predictions made.

Warp and twist and bend testing are recent additions to the test regimen and reflect a real concern over the trend in industry to use thinner boards. Here, compliance becomes a true necessity because the silicon chip or its solder joints are at a high risk of breaking during box assembly or due to mechanical shock. Underfill provides some protection, but a compliant layer is an advantage (Fig. 5.17).

This type of stress can have a profound influence on the reliability of a CSP. Compliance in all three dimensions is required to compensate for all the various thermal and mechanical effects that can damage the package or its leads.

5.7.3 Reliability Test Results

Tessera and others have performed extensive reliability tests on the package. As with most CSPs, package-level reliability is excellent. However, the concern of most OEMs is the performance of the package on the board. Figure 5.18 shows recent reliability results. *Note:* t_0 correction was applied to the data from test condition C (−65 to 150°C) that resulted in significant improvement in the goodness of fit (r^2) from 0.931 to 0.989. The Weibull ordinate scale and the characteristic life (*Eta*) are plotted in the t_0 domain. The actual characteristic life (*Eta*) of data from test condition C (−65 to 150°C) is 2857. Data for the chart are included in Table 5.14.

5.7.4 Failure Mechanisms

While the μBGA CSP is an extremely robust package, it is prone nevertheless to certain types of wearout failure. The package works as designed, and no solder-joint failures have ever been cited as a point of wearout failure. Figure 5.19 provides a finite-element model showing the efficacy of the strain relief provided by the compliant layer. (*Note:* Some brittle failures of the solder joints were

TABLE 5.12 JEDEC Moisture Sensitivity Requirements

JEDEC level	Description	Out-of-bag floor life allowed before assembly
1	Not moisture sensitive	Unlimited at 30°C and 85% relative humidity
2	Limited moisture sensitivity	1 year at 30°C and 60% relative humidity
3	Moisture sensitive	1 week at 30°C and 60% relative humidity
4	Highly moisture sensitive	72 h at 30°C and 60% relative humidity
5	Extremely moisture sensitive	48 h at 30°C and 60% relative humidity
6	Bake before use	6 h at 30°C and 60% relative humidity

TABLE 5.13 Sample of Reliability Requirements Testing for CSPs

Test required	Conditions	Duration
High-temperature storage	150°C	1000 h
Precondition thermal cycling, condition *C*	−65–150°C	1000 cycles
Condition *B* cycling (only if condition *C* fails)	−55–125°C	1000 cycles
Thermal shock, condition *C*	−65–150°C	1000 cycles
Steam pressure test	121°C at 2 atm	168 h
Precondition temperature and humidity biased	85°C and 85%	1000 h
HAST	130°C and 85%	80 h
Mechanical shock testing	1500-*g* peak	—
Bend/bow (bidirectional)	1.0-mm deflection	2.5 million cycles
Twist (0.25-mm-thick PWB)	5° and 9° twist	100,000 cycles

Source Intel.

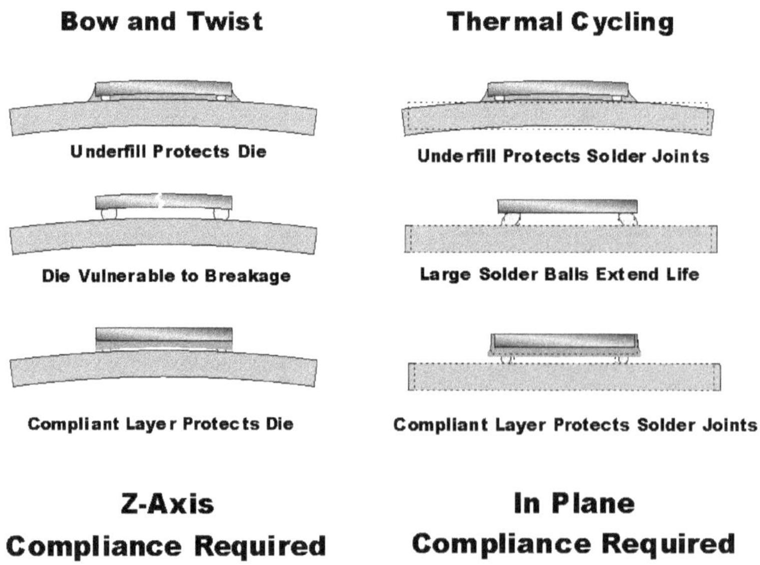

FIGURE 5.17 Warp and twist.

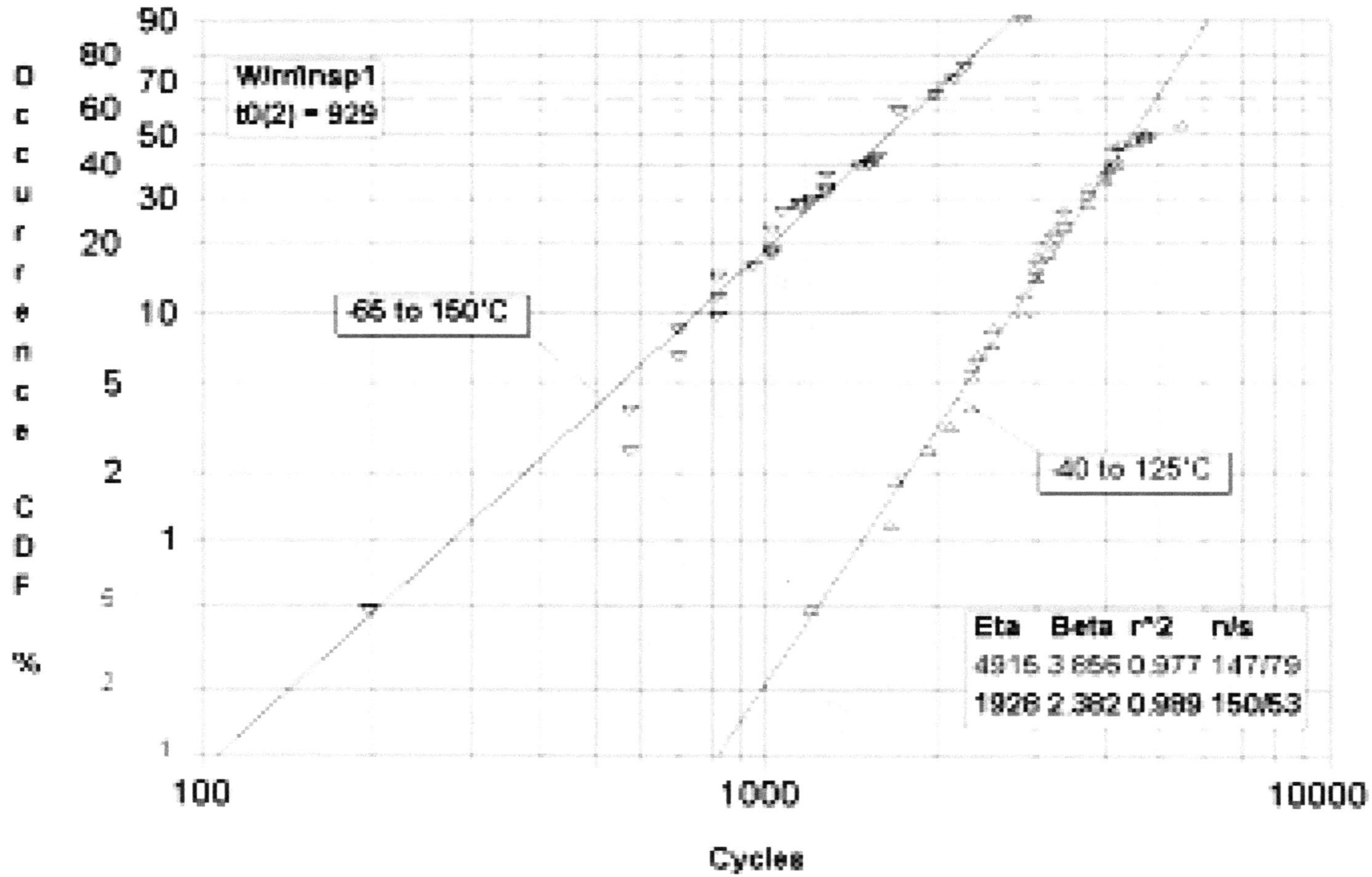

FIGURE 5.18 Example of a Weibull analysis of a μBGA CSP. Test conditions: −40 to 125°C on board.

TABLE 5.14 Data for Fig. 5.18

Run	Sample size	Fails/test point
1	30	0/2000, 2/2500, 5/3000, 11/3200
2	28	0/2000, 1/2500, 6/3000, 9/3500, 11/4000, 14/4100
3	29	0/1500, 1/2000, 3/3000, 9/3500, 12/4000, 13/4500, 14/4900
4	30	0/1000, 1/1500, 3/2500, 4/3000, 5/3500, 10/4000, 15/4500, 16/4700
5	30	0/1500, 2/2000, 3/2500, 5/3000, 7/4000, 11/4500, 12/5000, 13/5400
	147	
		Test conditions: −65 to 150°C
1	30	0/1000, 4/1500, 11/2000, 17/2400
2	30	0/1000, 2/1500, 9/2000, 17/2200
3	30	0/1500, 14/2000, 17/2400
4	30	0/2000, 3/2500, 7/3000, 14/3500, 24/3700
5	30	0/1500, 6/2000, 7/2500, 22/3000
	150	

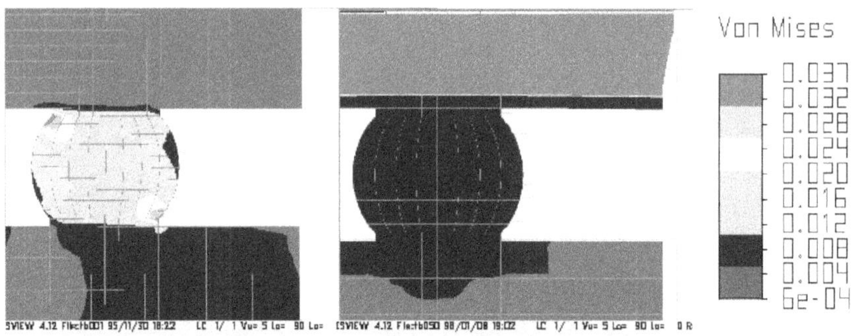

FIGURE 5.19 Solder joints on area array packages are the most vulnerable elements of the package in board-level reliability. It is desirable for the package structure to protect the solder joints by having the package absorb the strain of CTE mismatch rather than the solder balls, as shown in the right and left finite-element models, respectively. (*Courtesy Pacific Consultants LLC, Mountain View, CA.*)

reported on early versions, where gold thickness in the area of the solder joints far exceeded recommended limits.) The point of failure thus turns out to be the bond lead, which was designed to work in concert with the elastomer to protect the solder joint.

Lead failure is a classic wearout failure based on local work hardening. While the package and lead are designed to allow the elastomer to distribute the strain over the length of the lead, failure most often occurs at the heel of the lead near the bond pad. This location is the one most often predicted by finite-element modeling. However midspan breaks in the lead also have been discovered in a later test vehicle package (the TV-74). Figure 5.20 shows scanning electron micrographs of cracks in the package leads found in failure analysis. The result of this discovery was a full-scale review of the package concept and its mechanisms of failure.

Every failure that occurs in a μBGA package is driven by both the material used in its construction and the geometry of its features. There is thus a need to understand the primary factor that leads to a specific type of failure. Figure 5.21 illustrates the relationship between failure modes and the primary factors of material and geometry based on failure analysis observations.

The CTE mismatch between die and flex-circuit tape imposes stress on the leads during the thermal cycling. Depending on the geometry of the leads, the peak stress (or strain) will be on the heel, midspan, or shoulder. Therefore, lead heel break and midspan break can be classified as

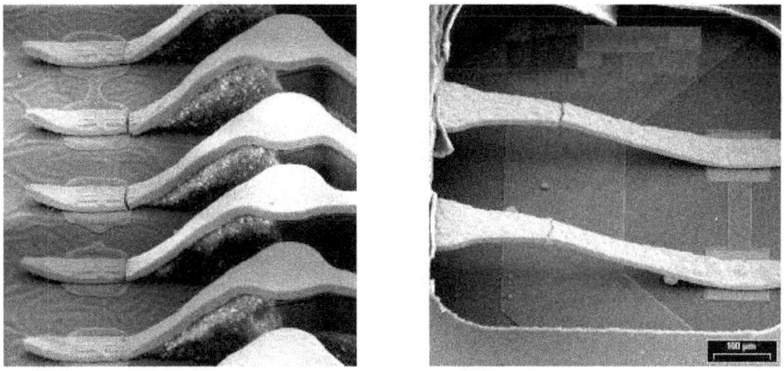

FIGURE 5.20 Lead heel break in TV-46 (*left*) and midspan break in TV-74 (*right*).

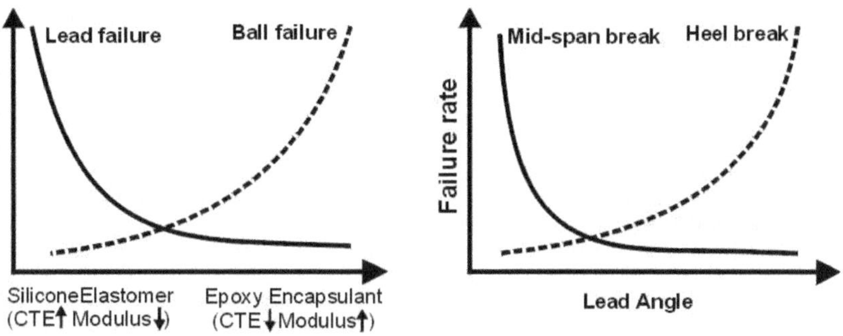

FIGURE 5.21 Failure mechanisms for the μBGA package can be changed by altering the variables of materials and design.

geometry-driven failure modes, whereas heel breaks are mainly due to fatigue from thermal cycling. The stress concentration at the heel can be minimized by reducing the lead angle, as illustrated in Fig. 5.21 (*right*). However, if the lead angle is reduced significantly, midspan break will become the major failure mode.

The failure mechanism of midspan break is not the same as that of heel break. With the reduced lead angle, the shape of the lead is not representative of the S shape typically sought. The actual lead shape more closely resembles a straight line. This condition offers no extra length to handle the CTE mismatch between die and tape. The result is a midspan break, which can be explained by the combination of tensile stress during the cooling half of a temperature cycle and lead buckling during the heating portion of a thermal cycle. The act of increasing the lead angle provides the potential for having more lead slack, thus avoiding the excessive tensile stress encountered during the cooling step. The higher lead angle also changes the lead deformation mode from buckling to bending under the heating step, resulting in reduced stress concentration at the midspan.

5.8 SUMMARY

The μBGA CSP is a highly reliable, low-cost, high-performance IC package. Moreover, it is capable of providing excellent electrical and thermal performance. Advances in manufacturing technology and refinements in package design are resulting in steady improvements in terms of both cost and performance. The assembler community is mastering the manufacturing technologies, and cost is expected to continue to decline. Next generation processes such as WAVE are in full-scale development and are expected to deliver on the promise of flip chip technology by providing a highly manufacturable device that provides standards and ease of assembly at highly productive rates.

Circuit card assemblers have reported that the μBGA package can be placed at rates of 20,000 units per hour with very few defects. Moreover, the field reliability data are extremely good because the compliance of the package protects the solder joints from failure. As cost targets are met and system designer and user demand continue to grow, the μBGA CSP is expected to play an increasingly important role in the future of electronic packaging over the full spectrum of products and pin counts.

5.9 REFERENCES

1. Fjelstad, J., "The Evolution of Area Array Packaging from BGA to CSP," in *Proceedings of SEMICON Europa*, Geneva, Switzerland, April 1998.
2. Kim, Y. G., "Low Cost and High Reliability Wafer Level CSP Technology," in *Proceedings of the IEEE Electronic Components and Technology Conference*, San Diego, CA, 1999.
3. Fjelstad, J., "Strategies for Creating Compliant IC Packages at Near Chip Size," in *Proceedings of ASME, INTERPAK,* held in Kauai, HI, 1999.
4. Johnson, H., and Graham, M., *High Speed Digital Design: A Handbook of Black Magic,* Prentice-Hall, Upper Saddle River, NJ, 1993, pp. 7–8.
5. Bogatin, E., "Electrical Principles of Packaging and Interconnects for High-Speed Digital Systems, Course 3: Creating Circuit Models with Approximations," course notes, 1994, pp. 2–21.
6. Johnson, H., and Graham, M., *High Speed Digital Design: A Handbook of Black Magic,* Prentice-Hall, Upper Saddle River, NJ, 1993, pp. 257–258.
7. Bogatin, E., "Electrical Principles of Packaging and Interconnects for High-Speed Digital Systems, Course 3: Creating Circuit Models with Approximations," course notes, 1994, pp. 4–9.
8. "Design Guidelines for Electronic Packaging Utilizing High-Speed Techniques," IPC-D-317A, Institute for Interconnecting and Packaging Electronic Circuits, Northbrook, IL, Sec. 5.7 to 5.8.3, 1995.
9. Lau, J., Wong, C. P., Prince, J., and Nakayama, W., *Electronic Packaging: Design, Materials, Process, and Reliability,* McGraw-Hill, New York, 1998, pp. 77–80.
10. Loo, M., and Gilleo, K., "Area Array Chip Carrier: SMT Package for Known Good Die," in *Proceedings of ISHM,* 1993.

ACKNOWLEDGMENTS

We wish to acknowledge the assistance of our many colleagues in the preparation of this material. Special recognition is due to Michael Perry of Pacific Consultants LLC for his contributions in the electrical and finite-element modeling. Also, recognition is due Max Nguyen and Anthony Faraci of Tessera for their work on and contributions to the reliability data and electrical modeling, respectively.

CHAPTER 6
FLIP CHIP TECHNOLOGY

Ken Gilleo
Cookson Electronics

6.0 INTRODUCTION

6.0.1 Basics

Flip chip (FC) is also called *direct chip attach* (DCA), perhaps a more appropriate term. The chip is flipped over so that the attachment side with the connection pads faces downward, giving the process its attractive name. However, it is really the method of connection that distinguishes this technology from all others. The interconnection is a joint without any wires. All other methods, including most chip-scale packages (CSPs), use wires between the chip pads and the substrate or package. There are other connection schemes, such as tape automated bonding (TAB), that connect to a circuit board with the chip face down, or "flipped," but the TAB package incorporates "wires." Thus, while TAB can have a flipped orientation, it can never have a DCA connection.

The concept of a direct chip connection to substrate adds significant benefits while also creating challenges for both assembly and the substrate. FC can be viewed as a nearly "packageless" interconnect system. The elimination of the package both adds and subtracts attributes. The package nearly always adds size and weight. Some near-chip-sized CSPs significantly increase thickness. The package adds protection and ease of handling and aids in standardization. We can see that an understanding of the package, with its penalties and benefits, is a requisite for evaluating the flip chip and other systems.

The Package. The package may add over a dozen features, but only a few are critical. We can think of the package as a structure consisting of a semiconductor device, a first-level interconnect system, a wiring structure, a second-level interconnection platform, and an enclosure that protects the system and provides the mechanical platform for the sublevel. The two critical attributes provided by most packages are environmental protection and ***geometric/mechanical translation.*** In other words, the package converts the extremely dense connection dimensions of the integrated circuit (IC) pads to the larger geometries typically used for the printed wiring board (PWB). The other important translation is mechanical, and this introduces compatibility between the ***inorganic*** world of the IC and the ***organic*** domain of most PWBs. The inorganic IC characteristics include very high modulus and low thermal expansion. The PWB is not nearly as stiff or dimensionally stable, and the coefficient of thermal expansion (CTE) is almost an order of magnitude higher than that of the chip. The package ***bridges*** together these two vastly different worlds. Elimination of the package is not a trivial matter and can require somewhat specialized circuit boards with the addition of added steps and new materials such as underfill. Figure 6.1 summarizes packaging attributes.

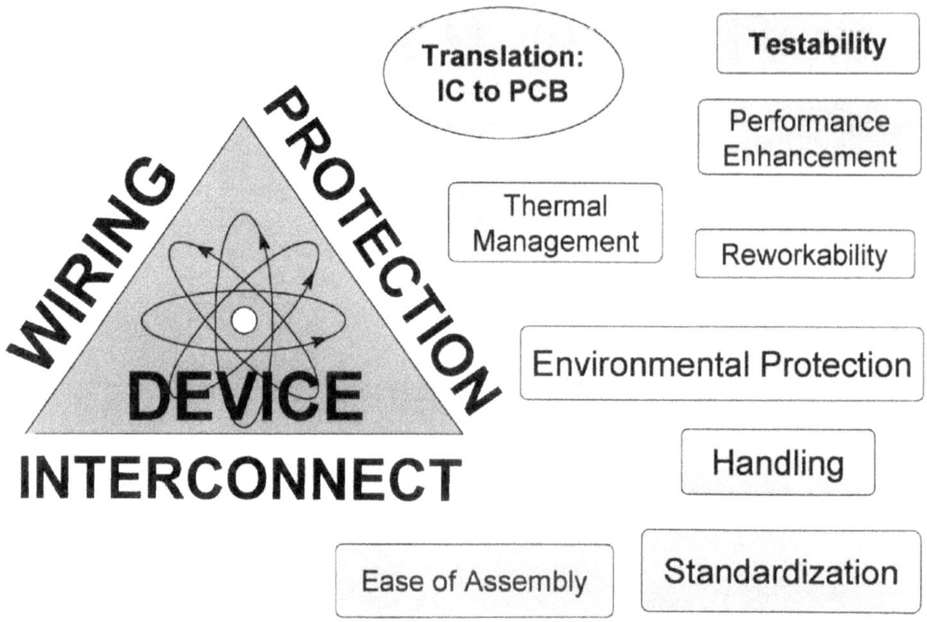

FIGURE 6.1 The package.

6.0.2 History

DCA was developed in the 1960s during the same time period that wire bonding was being established. Although a few early semiconductor manufacturers, such as Fairchild, explored the DCA concept, it was IBM who reduced it to a workable technology and a major commercial success. This was to be expected because IBM was *the* computer company, and mainframe computers were constantly pushing the envelope of performance. The FC benefits of higher performance, high I/O capability, and extreme density were just what the powerful mainframes needed. The FC also allowed heat to be removed efficiently, a feature important for computers but not for much else at that time.

The first DCA development did not even involve the IC, which was still evolving into a real product in the 1960s. Initial packaging work was done on transistors, which were being used in computers ever since IBM had made the pivotal decision to abandon vacuum tube logic for the new, but risky solid-state technology. IBM would need to move away from feed-through assembly if the room-sized mainframes were to ever get down to a manageable form factor. Therefore, much effort was directed toward directly bonding transistors to the substrate with a compact packing arrangement.

The first DCA product was the solid logic technology (SLT), a three-bump ball grid array (BGA) transistor. Earlier DCA work had involved direct soldering to planar pads on the transistor, but height, or standoff control, was a serious problem. The proposed solution was to use a nonfusible metal ball or sphere that would ensure good height control. Copper was the metal of choice because of its long use in electronics and excellent solderability. Copper microspheres were obtained from Alpha Metals, one of the specialty solder suppliers of the period. Attachment to the transistor was straightforward. The spheres were soldered to special finishes, now called *under-bump metallization* (UBM), using tin-lead alloy. High-melting alloy eventually was chosen to allow board assembly without sphere detachment from the device. Copper balls initially were supplied in bare copper, but later they were solder coated. The SLT is shown in Fig. 6.2.

The SLT is a very intriguing package—and it is a true package if we compare the attributes with those in Table 6.1. The transistor is passivated, so no additional protection is needed. The interconnect structure is made of nonfusible metal that could allow package removal, a requirement for the

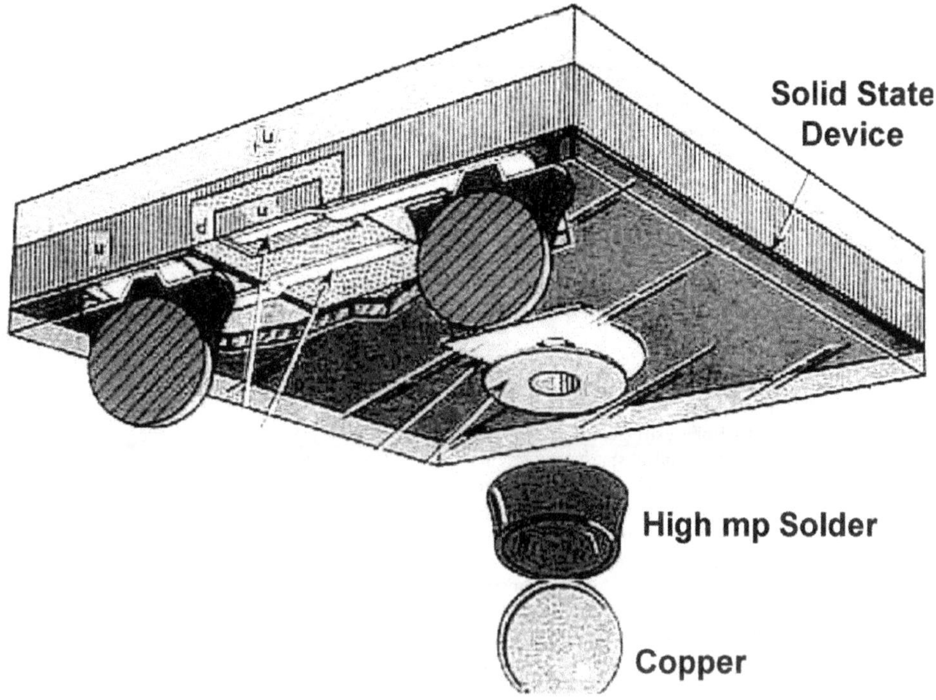

FIGURE 6.2 IBM SLT, the first FC?

true definition of packaging. Ostensibly, the SLT can be reworked, although IBM did not do this. However, look at its size. The SLT is chip size. The product certainly fits the dimensional requirements for a CSP. The SLT is probably the first commercial CSP, beating the AT&T-Allentown gold lead-frame product by less than a year. The AT&T package, with its cantilevered gold leads, was difficult to assemble and not easily reworkable and had limited success (Fig. 6.3). The SLT, the first practical FC, also adds a nice touch of irony if we look back on the CSP versus FC "battles" of the 1990s. The argument about which is better, CSP or FC, could not have been an issue 30 years ago because the FC was a CSP. The more recent FC versus CSP arguments may become moot again if FC becomes a CSP—*again*. And this is precisely what can be expected to happen, but more on the subject in later sections.

The final step in early FC development involved eliminating the discrete copper ball attachment and replacing it with a mass ball or bump formation method. Although a number of processes could be used, and many were tried, vacuum metallization won out. High-melting solder alloy could be deposited through a mask in an efficient wafer-level process. Although vacuum metallization equipment was rather expensive, the high throughput gave a good return on investment. Also, the solid-state electronics industry was much more comfortable with dry vacuum processes than with wet chemical methods such as plating that are only now becoming popular. The elimination of the non-fusible standoff would seem to bring back the height-control problem, but there was a simple solution. Height control was achieved by controlling the collapse of the bump. A high-lead alloy bump could be soldered to a circuit substrate using lower-melting solder while maintaining the height of the initial bump. Alternatively, the wet-out area on the wiring board could be designed precisely so that the amount of bump collapse could be dialed in. Not surprising, the system was called *controlled collapse chip connection* (C^4, and later, C4). The C4 process and its several variations are still used today.[1,2]

TABLE 6.1 Package Attributes

1.0 Platform/Chip Carrier
1.1 Mechanical support for electronic device(s)
1.2 Electrical ground as required
2.0 Interconnect
2.1 Means of electrical connection to device(s)
2.2 Signal paths with required transmission characteristics
2.3 Means of electrical and mechanical connection to PWB
2.4 Means of nonelectrical path when required, e.g., optical
2.5 Interface to PWB that allows removal and rework
2.6 Routing between device and PWB as required
2.7 Means of eliminating or counteracting thermomechanical stress
3.0 Protection
3.1 Mechnical, electrical, and environmental protection of device
3.2 Specific interior atmosphere as required, e.g., hermetic
4.0 Physical Structure
4.1 Robust structure sufficiently durable for assembly
4.2 Mechanical shape for handling and standardization
4.3 Surface for markings and nomenclature
4.4 Low-flammability materials and construction as required
4.5 Materials that meet environmental and work environment regulations
5.0 Thermal Management
5.1 Thermal path for heat removal as required
5.2 Integrated heat sink or spreader as required
6.0 Testability
6.1 Interconnect structure that permits electrical testing as required

6.1 UNDER-BUMP METALLIZATION (UBM)

The first step in preparing a chip for most of the bumping processes is to lay down a proper metallic base over the chip's bond pads. Most of the bumping processes are not compatible with the type and thickness of metal used in IC fabrication, which typically is very thin vacuum-deposited aluminum. Even newer copper metallization requires UBM. The UBM can be applied by standard vacuum coating techniques used in the semiconductor industry or by chemical plating. Vacuum deposition methods have been the most common, beginning with the very first developed by IBM over 30 years ago. More recently, maskless electroless plating has gained popularity because of simplicity that results in cost reduction and lower capital equipment requirements. A suitable UBM provides a barrier layer and serves as a tie coat, or "primer," for the bump that will follow. There is really no standard UBM, and many companies have their own proprietary composition of multiple metals based on many years of data accumulation. However, the electroless nickel UBM has a reasonable chance of emerging as a standard, especially among the new players in the bumping arena.

6.1.1 Problems with Aluminum

Most ICs use aluminum (Al) as the conductor layer to interconnect in-chip devices, although copper is starting to displace aluminum. Aluminum is a reasonably good conductor and is compatible with silicon materials. However, aluminum quickly forms a robust, chemically stable, and electrically insulative oxide that creates a connection challenge. Wire bonding with gold or aluminum wire produces a good, reliable connection because the aluminum oxide is removed in the bonding process. Aluminum is not soldered easily, especially because the layers used for ICs are very thin and easily damaged.

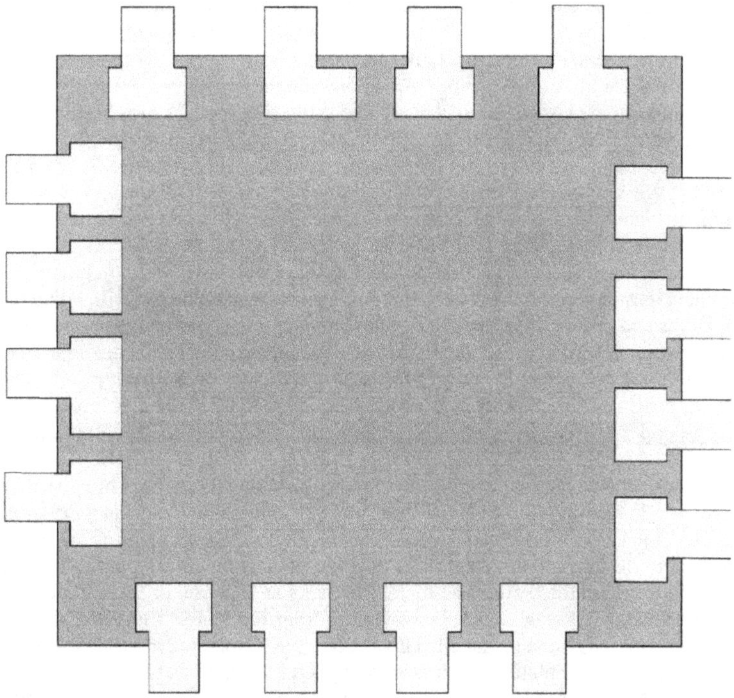

FIGURE 6.3 AT&T lead-frame CSP.

Only a few bumping or bump-forming processes can be used directly with aluminum pads. The most common is gold stud bumping,[3] a modification of gold wire bonding, although more recent work has been done with copper.[4] Virtually all other bumping methods require that a different metal finish be applied over the aluminum before creating the bump. Even conductive adhesives, also called *polymer bumping,* cannot be used directly on aluminum because the oxide insulates the conductive particles from the pure metal. Considerable research on adhesives, including those with penetrating particles, has not led to success with the bare metal. Although aluminum oxide can be removed by chemical or mechanical means, the oxide will "grow" back even with cured adhesive in place. The most common solution has been to apply a different metal or several metals to the aluminum pad prior to bumping. Since the metal finish ends up under the bump, it is referred to as *under-bump metallization* (UBM). Many different metals are used, but all serve as "primer" and sometimes as a barrier against metal diffusion.

6.1.2 Types of UBM

We can categorize UBM by either the type of metal or the method of application, but the latter is more meaningful. The early processes relied on vacuum deposition. The vacuum process is similar to the method used to put down aluminum conductor traces and pads. Metal is vaporized in a high vacuum and allowed to condense on the wafer through a patterning mask. Heating to a high temperature can vaporize the source metal, or bombardment with a stream of energetic accelerated ions or atoms can be used. The thermal evaporation used to vaporize the metal is often referred to as *metal vapor deposition* or *chemical vapor deposition* (CVD). Heat can be generated by applying an electric current to source containers called *boats* in a process called *resistive,* or *resistance, evaporation,* or *firing.* Since the heat flux and the evaporation rate fluctuate with this method, a controlled heat source, such as an electron beam, is more desirable for semiconductor work. Generally, only aluminum is vapor deposited by the IC industry because it is easily "boiled." Other methods are used to deposit copper, tungsten, and the more difficult metals.

Sputtering is a more manageable means of vacuum depositing. Virtually any metal can be sputtered, making the process ideal for UBM. The sputtering process uses high-energy gas ions to strike a metal target and dislodge the source, or target, material by transferring the high energy. Sputtering is a slower process than evaporation, but it can be controlled with extreme precision, virtually atom-by-atom. Even high-boiling metals, such as tungsten and chromium, can be sputtered because it is not a thermal energy process but rather an electrical/magnetic-powered atomic collision process. The basic principle is to energize a heavy inert gas, such as argon, and form plasma. The gaseous ions are then accelerated in electrical and magnetic fields when the common magnetron process is used. High-energy argon ions and neutralized atoms strike the metal target with enough energy to dislodge metal atoms and clusters. The metal particles can be directed at the substrate in a beam that produces a metal coating. Multiple metal targets can be used in a single chamber so that several different metal layers can be deposited in a single run. UBM may use a layer structure of two or more different metals to provide compatibility and good adhesion. Some of the common systems are aluminum-titanium-copper (Al/Ti/Cu), aluminum-nickel-copper (Al/Ni/Cu), and aluminum-palladium-copper (Al/Pd/Cu).

Metal also can be applied to aluminum pads by chemical plating. Three basic types of plating are recognized: (1) electrolytic, (2) electroless, and (3) immersion (a subcategory of electroless). All are being used for UBM. *Electrolytic plating,* also called *galvanic,* requires application of a dc current to the wafer and to a plating electrode. Positive metal ions in the plating bath are attracted to the cathode, in this case the wafer. The metal ion picks an electron or several, depending on its charge, to return to the metal state again. However, in order to connect the wafer die pads to an electrical source, a plating bus must be added. This is commonly done by vapor depositing an aluminum film on top of the wafer. A plating mask is first applied over the bus to define where plating will occur. Once the UBM is formed, the mask and bus are stripped away. The need to apply and remove a bus and mask adds steps and cost.

Electroless plating is a much simpler process that does not need current, but it is more limited.[5,6] Only a few electroless plating systems are compatible with aluminum. In fact, no commercial electroless plating systems are even compatible with aluminum. Fortunately, an immersion plating method can be used to produce a platable surface. Electroless immersion zinc plating can be applied directly to aluminum. Then electroless nickel can be applied over the zinc. *Immersion plating* is a type of electroless plating where a double ion-exchange reaction occurs between the plating bath and the metal being plated over.[7] In this case, zinc ions in solution are exchanged with aluminum atoms on the surface. The aluminum dissolves, and the zinc in solution deposits at the location vacated by the aluminum. Once the zinc plating bath has displaced the entire aluminum surface, the reaction stops. There is no more aluminum to drive the reaction, and thus it is said to be *self-limiting.* While immersion plating is self-limiting, nickel and most other electroless plating processes are *autocatalytic,* with the plating reaction continuing until the plating bath is exhausted.

These nonelectrolytic plating methods do not require a mask or bus, and this is an important benefit. One limitation is that only a few metals will plate electrolessly. Even though immersion and electroless plating generally do not require a mask, there is a thickness limit that is a direct result of the maskless method. A plating mask restricts plating in the plane of the wafer, the x and y dimensions, by providing sidewalls that restrict metal growth. Plating can only build up in the vertical or z direction. Plating buildup can continue up to the top surface of the mask. Without a mask, plating occurs in the x, y, and z directions. The deposit thus increases in width as it grows in height. Electroless plating therefore produces a mushroom shape as the reaction progresses. Continued plating will cause the metal at each pad site to grow outward and finally short out at adjacent pads. Since the UBM needs to be only a few hundred angstroms thick, this is not a problem.

The most popular newer UBM is electroless nickel with a zinc, or zincate, layer first applied to activate the nickel plating chemistry. In many ways, this is an ideal process because only chemical solutions are required. However, chemical composition control is very critical. Failure to control the zinc plating step, for example, will cause low adhesion. The nickel plating also might be deposited with high stress if the composition is not correct throughout the process. The stress will lead to adhesion loss. However, if the two plating steps are controlled correctly, an excellent nickel UBM will be produced. Since nickel will oxidize slowly, a gold flash, actually immersion gold plating, is commonly applied. The immersion gold is only about one atomic layer thick, and this is too insignificant to add cost or create soldering problems later. Nickel, with or without gold, is compatible with solder and many conductive adhesives. While some adhesives form stable junctions with a nickel surface, all perform well with a gold surface, and this is the preferred finish for polymer bumps.

6.1.3 The IC Transition to Copper

Several developments in the past few years have solved the numerous problems associated with using copper metal in place of aluminum as the IC interconnect metal. Copper is more conductive (about three times) than aluminum and allows higher frequencies to be used with smaller line widths. Many fabricators are converting to copper not just for speed but also for cost reduction. Thinner conductors allow closer spacing and smaller chips. The switch to copper allows many more dies per wafer, and this is where the savings come from.

Since copper is much more compatible with bump materials than aluminum, the transition to copper is expected to boost flip chip technology possibly by eliminating the UBM step. It remains to be seen what the final finish will be for copper ICs. The industry will continue to use wire bonding as the main interconnect method, so the final pad must be compatible with gold wire bonding. IBM has indicated that its copper chips will have aluminum as the final pad layer to accommodate wire bonding, and this may become standard practice. However, aluminum can be removed easily without affecting the copper underneath. Aluminum is an amphoteric metal that can be dissolved in both acid and base. Dilute caustic (sodium hydroxide) quickly removes aluminum. Many other reagents also can be used. Thus, even if the new copper chips come with an aluminum finish, a simple aqueous washing step will unveil the desired copper layer. In fact, the aluminum over copper would serve to protect the copper from oxidation. Gold over nickel also could be used on the copper pads similar to PWB common finishes. This too would be a very good surface for most bumps.

Conductive adhesives would receive a real boost in the switch to copper because none are compatible with aluminum. Some of the conductive adhesives form reasonably stable junctions with bare copper, especially those using an oxide-penetration mechanism. Even for those adhesives which are not suitable for bare copper, simple UBM methods could be used. Silver and other finishes can be applied to copper by electroless, maskless plating. The advent of copper-based chip interconnection metallurgy undoubtedly will simplify FC fabrication in the near future. It probably will be possible to directly bond the copper pads with conductive adhesives. This simple processing ability would have a great impact on cost and infrastructure issues by eliminating UBM and maybe the bumping step. Assemblers could run the entire FC preconditioning and bonding process.

6.2 BUMPING MATERIALS

There is some overlap between UBM, just covered, and bumps. Some UBM processes can be extended to produce complete bumps. UBM can be thought of as the very thin metal primer, or tie coat, on which the bump is deposited. The UBM will always have more conductive material added to increase thickness beyond the die passivation layer. Some UBM processes, such as electroless nickel, can be continued until a thickness is built up by many orders of magnitude to produce a bump that rises above the die passivation layer.

In theory, any conductor can be used for bumps, provided that it has sufficient strength and integrity to hold the chip to the substrate. Bump materials can be divided into fusible and nonfusible classes. The difference is quite important. Many fusible types, especially those of lower-melting alloys, provide the joining material for assembly. While it is desirable to have the joining material built into the bump, this can have a downside such as more difficult height control. A bump made totally of the same fusible alloy can totally collapse by spreading out on the connection surface if the substrate pads do not have a means to restrict the wetting area.

Nonfusible bumps, such as the copper spheres used in the SLT, solve the height-control problem. The penalty is obvious: Joining material must be added. This requires an added step and just one more place where things can go wrong if good control is not maintained. However, the nonfusible bump can add features that offset the added step cost. Proper design can permit rework. Higher standoff can improve reliability. The higher melting point of the bump can be ideal for flip-chip-in-package (FCIP), which will experience reflow temperature at least one more time. Thus each type has its advantages, and the end use often will determine which way to go.

6.2.1 Fusible Bumps

Fusible bumps are now the most widely used bump type because they can supply the joining material and are easy to make. However, some fusible bumps have a high enough melting temperature to allow use as nonfusible bumps. Thus my definition will get a little fuzzy unless the final assembly is considered.

High-Lead Solder. The original C4 process used high-melting tin-lead (Sn/Pb) solder such as 3/97. This is still a fusible alloy, although a reflow temperature as high as 350°C may be needed. This is a practical temperature for ceramic substrate, however. IBM and other mainframe computer makers have used ceramic substrate and continue to do so because of its unique attributes. One important reason is the high thermal conductivity. Some ceramic circuit materials are better thermal conductors than metals while serving as excellent dielectrics. The massive number of chips in a mainframe and the high board and system density suggest that ceramic materials will continue to be used here and that the high-melting-alloy bumps will see service into the foreseeable future.

Some of the ceramic FC BGAs also use high-lead bumps. This allows the package to be soldered to a board with eutectic solder without disturbing the FC-to-chip-carrier joints. The ceramic substrates, while more costly than organic substrates, give more planar surface and low moisture absorption. Low moisture absorption means that packages do not have to be stored in a dry atmosphere or predried prior to solder reflow. Moisture absorption, as seen in most organic BGAs, causes explosive evolution of steam and instant delamination, called *popcorning.* It also should be mentioned that the high-lead bumps are less prone to thermomechanical fatigue. All told, high-lead bumps serve a function, and their important niche uses will continue.

Stratified Bumps. One clever idea is to build a high-lead bump that is rich in tin near the top. This is also called the tin-cap process. Both IBM and Motorola have been active here. This kind of structure is said to provide the high standoff and favorable fatigue resistance of high-lead bumps while creating eutectic solder for the joint.

Eutectic Sn/Pb. Eutectic solder is the de facto standard of electronic assembly, so it is no wonder that this same material has become essentially the standard solder bump. First, a eutectic bump can be used as the source of joining material for assembly on organic PWBs. The industry has more or less geared itself to processing around the conditions required by eutectic solder. Next, eutectic tin-lead solder is readily available in a variety of forms, including fine-particle paste that can be used for bumping. Since the eutectic temperature is the natural equilibrium point of an alloy, it is the ideal region for processing. However, eutectic solder joints are prone to fatigue and will fracture eventually if subjected to continual stress.

Polymer Bumps, Thermoplastic. Polymers can be either fusible or nonfusible. Thermoplastics behave like solders in that they melt on heating and solidify on cooling in a process that can be repeated many times over. Thermosets, on the other hand, do not melt once they are set, or polymerized. Limited work has been done on thermoplastic-based conductive adhesives as bumping materials. The adhesive bump can be formed by applying an isotropic conductive adhesive (ICA) to the UBM or to a nonfusible metal bump. ICAs represent one class of conductive adhesives that are highly filled with metal and conduct equally in all planes. The other class of adhesive is the anisotropic conductive adhesive (ACA) with unidirectional electrical conductivity. The ACA is considered to be a joining material, not a bump-forming system, and will be covered in a later section.

Thermoplastics are polymerized solids, but they can be made into conductive pastes by dissolving them in polar solvents and adding conductive particles. The most common conductive particle is silver because of its stable conductive oxide surface, ability to be formed into the correct morphology, and a reasonable balance of cost and performance. Silver filler is typically a combination of small flakes and spherical particles ranging in size from 5 μm down to even submicron size. Loading for the ICAs is about 80 percent by weight, but excluding solvent. Once applied, the solvent can be evaporated to form a dry bump. The bump can be attached to substrate by heating to the softening or melting point. Figure 6.4 shows bump construction.

6.2.2 Nonfusible Bumps

Nonfusible bumps make up only part of the interconnect between IC and substrate. Joining material must be added to complete the connection. Therefore, the nonfusible bump produces a more complex joint made up of a different composition at each interface and possibly intermediate regions of both compositions including intermetallic compounds. However, the nonfusible bump can provide several benefits over fusible materials. Depending on hardness and other properties, the nonfusible bump may be tested and burned in. A nickel or copper bump can be probed, socketed, and burned in much more easily than a soft solder bump. Virtually all nonfusible bumps offer controlled standoff so that a minimum height can be maintained. This can be a very significant feature because too low a standoff will increase strain dramatically. A low clearance between chip and substrate can cause underfill voids or even make filling impossible.

Gold. Gold was used to bump chips even before it became popular for flip chips. The tape automated bonding (TAB) process uses a gold-to-aluminum connection metallurgy and requires a bump somewhere in the process. Gold bumps can be created on the IC pads or on the TAB chip carrier.[8] Methods developed to form gold bumps on aluminum IC pads for TAB also can be used for flip chips. Pure gold is used typically and can be applied directly to aluminum pads without UBM. The gold bumps are highly compatible with all conductive adhesives. However, soldering requires specialty alloys.[9] Eutectic tin-lead solder will leach or dissolve gold and form metal compounds called intermetallics that greatly reduce joint reliability because of high brittleness.

Although a few companies are promoting gold bumps and solder attachment with special alloys, most use conductive adhesives as the joining material. Gold is soft and malleable, allowing bumps

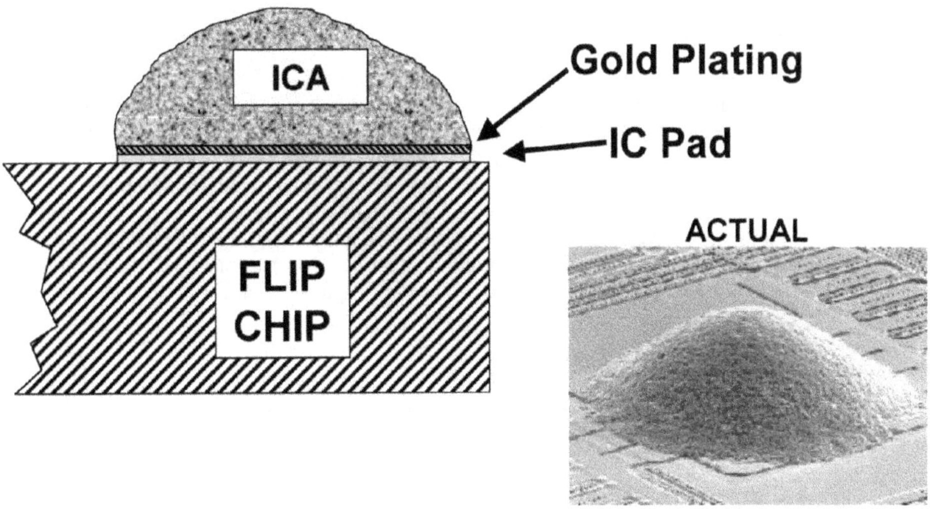

Courtesy of Polymer Flip Chip Inc.

FIGURE 6.4 Polymer bumps.

to be shaped easily. The type of joining material determines the shape requirement. ICAs work well with conical shapes, whereas flat bumps are more suitable for ACAs. I should note that unlike lead-based solders, which emit alpha particles that cause problems with many ICs, gold contains no radioactive isotopes. The use of nonemitting conductive adhesives therefore provides a system with no radioactive emissions, making it ideal for memory and other types of chips that are sensitive. Shaping will be covered under bumping processes below.

Nickel. Nickel is a very strong, hard, and high-melting metal that is used extensively in electronics. Nickel is soldered readily and also is compatible with some conductive adhesives. While nickel can be used as a UBM, it also can be formed into bumps. In most cases, the nickel serves the dual purpose of UBM and bump. The nickel plating process is simply continued until the thin UBM coating builds up to a bump. Once formed, the nickel bump usually is overplated with a thin coating of gold to prevent oxidation.

The nickel bump can be coated with solder to form a multiple-composition system. Although it can be argued that the nickel is then serving as a UBM, if there is substantial height, the nickel should be classified as a bump. The purpose of producing a bump instead of just UBM is to create a minimum standoff. At least one of the bumping processes produces shallow bumps because the bumps widen as they grow in height. The bump height may be less than 25 μm (1 mil), with the fusible portion making up most of the height.

Polymer, Thermoset. Polymer bumps usually are made with thermoset ICA materials that are stenciled or screen printed onto the UBM. A gold-finish UBM typically is required, although nickel can be used if the adhesive can form a stable junction. However, since a simple and inexpensive process is used to apply a flash of gold, bare nickel is seldom encountered. Silver-filled epoxies are by far the most common bumping material. Their conductivity is nearly the same as that of solder, and their mechanical strength is reasonable. Conductive adhesive joints still fatigue when thermomechanically stressed, although some give modestly better performance than solder bumps. Once cured, the conductive adhesive bump requires a joining material for assembly and this simple, more conductive adhesive paste. However, Fujitsu claims to make bumps with thermoset adhesive that are hardened but not fully cured (B stage). These bumps can be attached by applying heat and pressure. The B-staged thermoset bonds and then fully polymerizes.

6.3 BUMPING PROCESSES

This area of technology is still evolving, and by the time you read this, a few additional methods may have been announced. All bumps are made of metal or composites containing mostly metal. The bump may be deposited directly or as a material that is converted to a bump by a subsequent process such as solder reflow. The methods may be roughly grouped into discrete one-at-a-time singular and mass-deposition parallel processes. While mass deposition is most popular, discrete methods such as stud bumping are also enjoying success. The mass-deposition methods can be subdivided into physical material transfer (such as printing), chemical (such as plating), and vacuum deposition, although the last also can be classified as a mechanical-physical process.

6.3.1 Vacuum Deposition

Let's start with vacuum deposition, also called *vacuum plating,* because this was the most successful early process and still remains an important one today. The basics of vacuum deposition were covered earlier. A vacuum-deposited bump is always formed on the appropriate UBM. Virtually all metals can be vacuum deposited, as well as alloys, although it is more common to apply pure metals in sequence and then form the alloy by melting in a later step. Lower-boiling metals such as tin and lead can be evaporated easily in a vacuum, and this is the primary vacuum method for tin-lead alloys. The deposition time, and thus the amount, of each constituent can be varied to produce the desired alloy. The same vacuum chamber therefore can deposit 3/97 and 60/40 tin-lead just by changing the parameters.

A mask is used to define the deposition areas. A metal contact mask can be positioned over the wafer with openings that correspond to the IC pads. Masks are made from high-temperature, precisely imaged materials including titanium, tungsten, and molybdenum ("moly mask"). These masks are used repeatedly and cleaned as required. Metal deposits can be removed by melting or by chemically dissolving the bump material in reagents that do not affect the mask.

Integral masks, those applied directly to the wafer, also can be used. A photoimageable polymer is employed, although laser machining could be used with nonphotosensitive polymer films. The mask material is applied by spin coating and then imaged and developed in place. Once the bumps are formed, the mask is stripped away and cannot be used again. The integral mask can provide better registration and resolution because photolithography is used. Turnaround is quicker because a phototool is used. The process is also called *lift-off.*

Applying one metal and then the other(s) can produce alloys. The process can be alternated to provide more homogeneity, although this is not necessary because the bump is usually reflowed as part of bump manufacture and often reflowed again during assembly if it is a fusible type. A stratified bump with a low-melting tip can be produced by first applying the high-melting composition and then the lower-melting one. One process constructs a high-lead bump and ends by applying a tin cap that becomes a eutectic tin-lead composition during reflow. This approach provides an assured standoff height while allowing the bumps to be soldered at lower temperatures without adding solder pastes.

6.3.2 Plating

Chemical plating is a common industrial process whereby metal salts in solution are converted back to metal on the surface that is to be coated. The process is divided into *electrolytic,* where electric current is the controlling and driving force, and *electroless,* where chemical reactions supply the energy.

Electrolytic Plating. Most metals form at least one water-soluble salt, although the pH may need to be adjusted for optimal results. A metal salt consists of a metal cation with a positive electrical charge of 1, 2, 3, or higher corresponding to its chemical valence, the number of easily lost electrons. One or more anions balance the positive metal cation. The anion is a species of opposite charge that results in a salt that is electrically neutral. When a direct current (dc) is passed through the aqueous salt solution, positive metal cations are attracted to the negatively charged cathode. Acquiring the appropriate number of electrons to convert back to the metallic state neutralizes the cation. Since metal is being depleted from the solution, or plating bath, an electrode of the same metal typically is used for the positive anode, where metal ions lose electrons to form replenishing cations. Solder alloy can be plated, or the individual metals can be stratified and melted into solder.[10] However, it is more efficient to codeposit metals and form the alloy in situ.

Electroless Plating. Metal plating also can occur without the use of an electric current. A reduction-oxidation, or redox, reaction must occur for "automatic" electroless plating to result. Electroless plating can be further divided into autocatalytic, where any thickness is possible, and immersion, where plating stops when all the base metal is covered up. Both types of plating are used in electronics and in bumping.

Electroless nickel plating, covered earlier, became very popular in the mid-1990s as both a UBM and a bumping method. Nickel systems will plate onto a number of surfaces, but aluminum is not one of them. Using zinc immersion plating solves the problem of plating onto the still-common aluminum pads. Zinc salts will react with aluminum to convert that metal to a salt while simultaneously being reduced to metallic zinc. This is an immersion plating reaction where the zinc cation is reduced to metal, and the aluminum is oxidized to a salt, hence the term *redox.* Once the zinc has been deposited by immersing the wafer in the zincate solution, we are ready for nickel plating. It should be noted that the zinc plating, or zincate treatment, is a critical process, with cleaning and multiple treatments often required for optimal results. In a sense, the zinc layer can be thought of as the UBM for the nickel bump.

Electroless nickel plating can now be deposited over the zinc. The nickel will continue to plate out until the bath is depleted because nickel is a catalyst for its own plating reaction (autocatalytic). Time and temperature determine the plating thickness, so precise control is possible. As we saw earlier, a UBM deposit only requires that sufficient nickel be applied to serve as a base for the bump that typically is solder. A nickel bump is many orders of magnitude thicker because it must extend well beyond the die passivation layer that is several microns thick.

The electroless nickel process can be used to form bumps by increasing the plating time until the height is approximately 2 to 40 μm. However, unless a plating mask is applied, and this is not usually done, the nickel will plate upward and outward, forming a mushroom-shaped bump, as shown in Fig. 6.5. Continued plating eventually will cause the individual bumps to touch and short out. Pitch and the spacing requirement between bumps therefore limit the useful plating thickness or bump height. Although this is a restriction for high-density FCs, many classes of FCs can use the process at substantial cost savings. The flat bumps, especially with gold flash plating, are ideal for use with anisotropic conductive adhesives but also can be attached with solder paste. A common practice involves applying a very thin layer of gold over the nickel as the final step. This prevents oxidation, and the small amount of gold does not cause any problems with solder embrittlement. The gold plating process is immersion gold that stops when all the nickel is covered up.

6.3.3 Printing/Stenciling

Solder-paste printing on wafers is a very efficient and low-cost method of forming bumps. Although screen printing can be used for low-density chips, stencil printing is superior and used exclusively.[11] The advances in stencil making over the past several years have enabled the solder-paste bumping process, but stencil feature size capability is still the limiting factor for the process. Stencils made by the highly precise electroforming method (uses electroplating to form the stencil) are considered optimum, although laser-cut stencils also have been used. Specialized solder pastes give best results.[12] A solder-coated process using metal salts that are thermally decomposed to form solder also has been reported.[13]

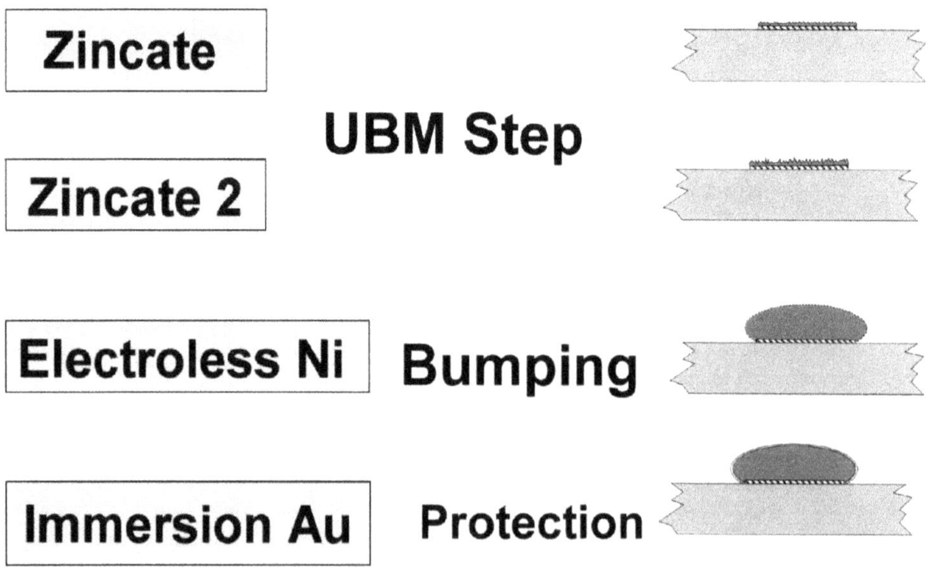

FIGURE 6.5 Electroless nickel bump process.

The precision stencil is aligned to the wafer pads, and fine solder paste, type 4 or 5, is applied after the appropriate UBM has been laid down. The paste is then reflowed to attach the solder to the UBM and form the semispherical bumps. Several groups have defined the process parameters and reported successful commercialization of the process. Solder-paste stenciling probably has become the most popular FC bumping method for all but the high-density chips.

There is one other modification of the solder-paste stencil process worth mentioning—integral mask. The stencil is created on the wafer using photoimagable resist. Only the pad areas are opened. Solder paste is applied to the openings using a doctor blade. The paste is reflowed, and the mask is stripped away. The unmasked bumps may be reflowed a second time to produce the desired semispherical shape. This method allows higher density bumping and is thought to be used by Flip Chip Technologies, Inc. Patents may be pending or have been issued by now, and anyone contemplating using integral mask deposition should check the patent literature.

6.3.4 Metal Fluid Jetting

A technique has been developed in which a microdrop of liquid metal is propelled to locations on a substrate with extreme precision.[14] The hot metal can adhere to UBM surfaces to create bumps. Reflow will provide a stronger bond and round out the bump. Equipment is now available that can be used to bump wafers at speeds of 1000 or more drops per second while still maintaining high precision. One style of equipment from MPM/Speedline moves substrate while a piezo "gun" fires the solder spheres. This *shoot-on-the-fly method* provides a good balance of productivity and precision and bump sizes down to at least 75 μm. However, this is a serial process and probably not optimum for very high I/O dies. Figure 6.6 shows "jetted" bumps.

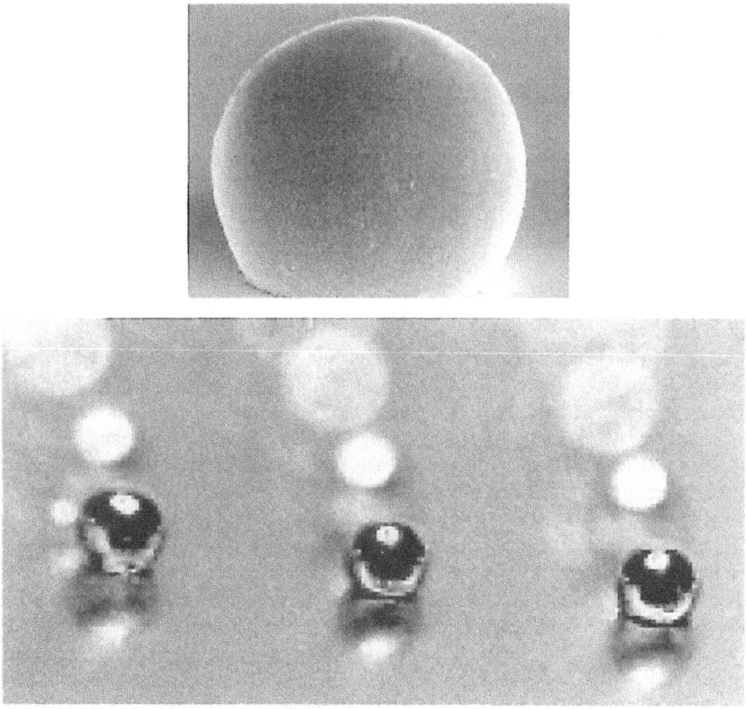

Courtesy of NIST

FIGURE 6.6 Solder jetted bumps.

6.3.5 Mechanical

There is only one important mechanical bumping method, and it is gold-stud bumping, a rediscovered process that dates back to the 1980s. The process is a modification of gold-ball bonding and has been around for a dozen or more years but is still covered by a patent.[15] Earlier use was strictly for prototyping, but several companies are using the method in production because of a number of unique features.[16] The gold-ball bond is formed, and the wire is broken close to the ball instead of making the loop. There are several important attributes to the stud-bump method:

1. No UBM is required.
2. The process is completely mechanical; no chemicals are used.
3. The process is practical for single dies as well as wafers.
4. It is lead-free, with no environmental issue or alpha-particle emission concerns.

This simple bumping method can be used without substantial investment or permits from regulatory agencies. However, there are a few significant limitations that reduce widespread use. First, a joining material must be added, and common Sn/Pb eutectic solder cannot be used because gold intermetallic compounds will form. Indium and other special alloys work, but at a higher cost and with certain performance deficiencies. Conductive adhesives have become the most common material for use with stud bumps, and products, such as memory chips, have gone with this approach. Methods have been worked out to increase stud-bump height and to control shape. A double bump can be used to increase height. A secondary shaping operation can be applied to form a conical bump that is preferred for use with adhesive pastes. Figure 6.7 shows a conical gold-bump profile. The joint is formed using isotropic conductive adhesive (ICA).

6.4 JOINING MATERIALS AND AGENTS

Ancillary joining agents such as flux may be required to form FC joints when the joining material is provided by a bump made of solder. However, several types of nonfusible and high-melting bumps are used, including high-lead solders that require joining materials such as eutectic solder alloys or conductive adhesives. This section will describe the most common materials in use today for joining FCs to substrate.

6.4.1 Flux

The presence of oxide on bumps or on the wiring board pads interferes with soldering and the formation of good metallurgical junctions. The simplest remedy is to convert the metal oxides back into pure, easily wet metal. Flux is a chemical compound that contains a reducing agent to convert metal oxides to metals just prior to solder joint formation. While flux can be incorporated into solder paste, this section will deal with stand-alone fluxes. The most common flux agent is organic acid, and nearly all fluxes for electronics use this chemistry. Once the flux agent does its job, it must be either removed or deactivated to prevent corrosion later on. The presence of acid in contact with metal can cause a host of problems later, especially in high-humidity environments. Flux can have other functions, such as holding the FC in place as the assembly moves into the reflow oven.

The FC presents special challenges to flux chemists, however, and many good materials for standard SMT cannot be used. Flux residue cannot be removed easily from under the FC because the gap can be 25 μm or less. "No clean" fluxes have become popular for FCs, and many advancements have occurred. However, if flux residue is left under the FC, it should be compatible with underfill. Many "no clean" fluxes were found to be very incompatible with underfills. Since the underfill must form a very strong bond to both the wiring board and the bottom of the chip, any flux that reduces the bond strength or performance of the underfill would degrade reliability.

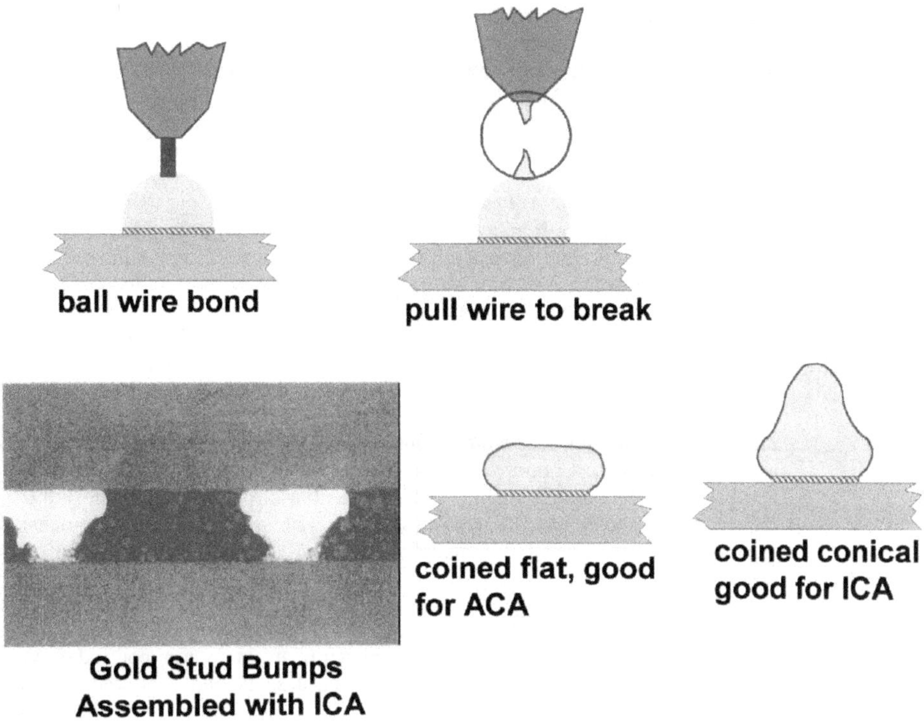

FIGURE 6.7 Gold stud bumps.

In the early 1990s, Alpha Metals hit on the idea of using underfill ingredients to make flux. Since nearly all the popular underfills are based on epoxy resins, fluxes with similar chemistries were investigated and developed. Active flux agents were selected that could polymerize with the epoxy after reducing the oxides. The most useful class of flux/hardeners is multifunctional organic carboxylic acids. A series of FC epoxy-based fluxes have been commercialized, and the concept has received broad patent coverage. The flux residue becomes a hard, strongly adhering epoxy thermoset after exposure to solder reflow temperatures. Underfill will react chemically with the flux surface so that no delamination will occur. The epoxy fluxes can be made in a solventless form over a wide range of viscosities with different tackiness values. These are the preferred fluxes for FCs using eutectic solder, although several more standard fluxes have been found to work as well.

6.4.2 Solder Paste

Solder paste is used for nonfusible bumps such as high-lead, high-melt-point solders and plated metals such as nickel. Although alloys such as 3/97 tin-lead can be melted and soldered, the reflow temperature of up to 350°C is too high for anything but ceramic substrate. A more common practice is to apply solder paste to the wiring board, place the FC, and reflow the solder paste in a process similar to SMT. Solder pastes for FC assembly should be of a finer particle class such as type 4 or 5, especially for fine-pitch work. The paste also must have enough tack to hold the FC in place during transport to the oven. The paste ingredients should be compatible with underfill. Epoxy-based flux technology has been applied to solder pastes.

6.4.3 Conductive Adhesives

Conductive adhesives offer several benefits for FC assembly. Both anisotropic (materials with unidirectional conductivity) and isotropic (equal conductivity in all directions) conductive adhesives can be used. In fact, even nonconductive adhesives have had success when used as "mechanical springs" to force bumps against pads. The limitations of adhesives when used for standard SMT, such as lower strength, are mitigated when they are used for FCs, making them an important joining material.

Anisotropic Conductive Adhesives (ACAs). First glance suggests that there is magic in an adhesive that only conducts in one direction, the *z*, or vertical, plane. A closer look shows that the *z*-axis adhesives are rather ordinary. The materials consist of an insulative binder that can bond to the chip and substrate. The conductors are small particles, preferably spherical, that are dispersed in the binder at a low enough loading so that interparticle contact cannot occur. In one sense, an ACA is an ordinary conductive adhesive without enough conductive filler. Electrical connections occur when the conductive particles are trapped between opposing bumps and circuit pads, as shown in Fig. 6.8.

One can see from Fig. 6.8 that ACAs produce a pressure-contact interconnect instead of a metallurgical junction like solder. This means that pressure must be maintained for reliability, and this is the key to good performance. ACAs only work well with highly planar systems or a substrate that is comformable, such as flexible circuits.[17] Fortunately, FCs are very planar, allowing ACAs to be used. Since ACAs are capable of connecting high-density structures, they are a good fit for FCs and have achieved success, especially for chip-on-glass (COG) used with flat-panel displays.

Modern ACAs use randomly dispersed conductive microspheres. The list of particles that have been used is extensive and is found in Table 6.2. Early ACAs used solid conductors, such as nickel, silver, or gold particles, but coated nonconductors are more common today. Metal may be plated onto glass, plastic, or other materials using electroless methods well known to the industry. A very popular system is gold plate over nickel plated onto elastomeric microspheres. The plastic spheres are cleaned, activated, and then plated with electroless nickel and a final flash of gold by immersion

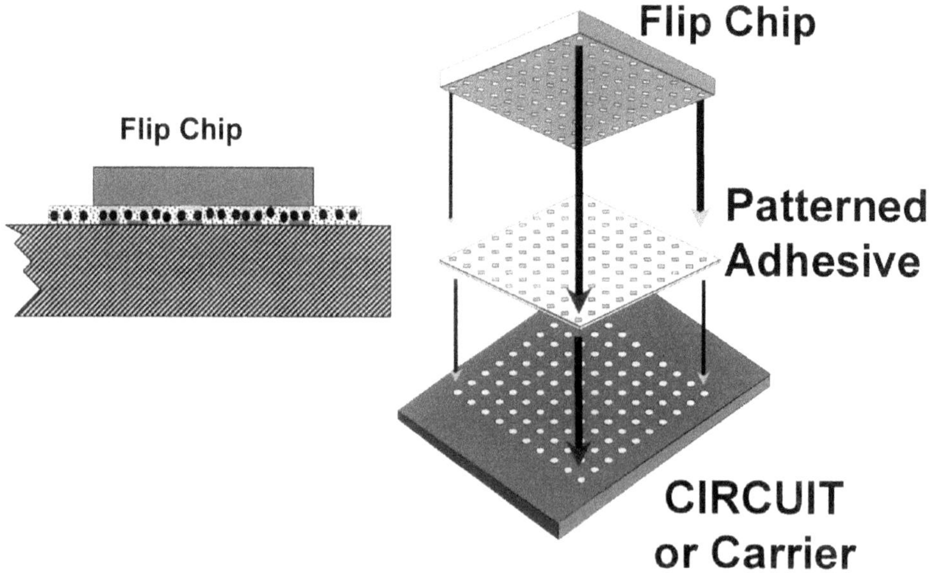

FIGURE 6.8 ACA junction.

plating to prevent oxidation. A coating of pure gold over plastic gives better performance, but electroless gold plating is more difficult and expensive.

The conductive particles are dispersed in a polymer adhesive at only a few weight percents. The adhesive can be a thermoplastic, a thermoset, or a blend of both. The thermoplastic ACAs have extraordinary shelf life at room temperature but are limited to an operating temperature of 100°C or slightly higher. Thermosets, with or without added thermoplastic, are more common and can pass aging and cycling up to 125°C. Their limitations are storage life, higher bonding temperature/pressure, and longer assembly times. Products are commonly available as films, including B-staged thermosets, but pastes are also sold. Finally, it should be noted that the ACA binder also serves as underfill, thus eliminating this bothersome step.

Isotropic Conductive Adhesives (ICAs). ICAs are exemplified by the commonly available silver-filled epoxies used for die-attach and solderless SMT assembly. These materials are highly loaded with silver filler so that the intimate contact produces good conductivity in all directions. Silver is used as the conductor because of its optimal balance of performance versus cost. Silver oxide, unlike the oxides of copper and the common base metals, is conductive. Although nonoxidizing metals such as gold can be used, silver provides equal performance at a cost savings. Silver is quite ductile and can be milled into the right flake morphology to produce high conductivity and the required rheology. Flake, because it can form an overlapping matrix, must be used for good performance. Powder alone would result in lower conductivity and a general instability. Silver flake also provides thixotropy, which allows the dispensed adhesive to stay in place. Figure 6.9 shows the model for ICAs. Special adhesives have been developed for FCs using finer silver fillers and additional milling to break up any agglomerate that would cause an open in fine-pitch assembly. The junction resistance will be higher than solder for most materials but not enough to be a concern. However, thermal conductivity is only about one-tenth that of solder, and this could be a problem for high-thermal-output devices.

Adhesives, Nonconductive. First, nonconductive adhesive is not used as part of the bump-to-board junction. The adhesive serves only as a mechanical link, pulling the chip to the substrate. A pressure contact must be created between the bump and circuit pad. Gold bumps and pads generally are used because a gold-to-gold pressure contact is extremely reliable. The adhesive should provide tensional forces, acting like a spring. This has been accomplished by using polymers that shrink on cure. While thermosets have been used, ultraviolet-cured systems have found success in chip-on-glass (COG) applications because the "light" can be transmitted through the flat-panel display from the opposite side.

Some data suggest that the tension of the adhesive relaxes after several years, causing opens. The proposed solution is to create compliant bumps from elastomers that are vacuum coated with metal,[18] but this concept has not gained commercial significance. Gold contacts tend to fuse when under pressure for extended periods, so there may not really be a problem.

TABLE 6.2 Adhesive Fillers

Conductor	Comment
Silver	Used early; can cause electromigration
Nickel	Oxidizes; may be unstable
Gold	Expensive; seldom used
Gold-over-nickel	Cost-effective, hard, nonelastic
Silver-on-glass	Low cost, hard, nonelastic
Gold/nickel-on-plastic	De facto standard; good performance
Gold-on-plastic	Appears to be the ultimate in performance
Solder	Has been used in multilayer circuit applications
Carbon fiber	Low cost, higher resistance

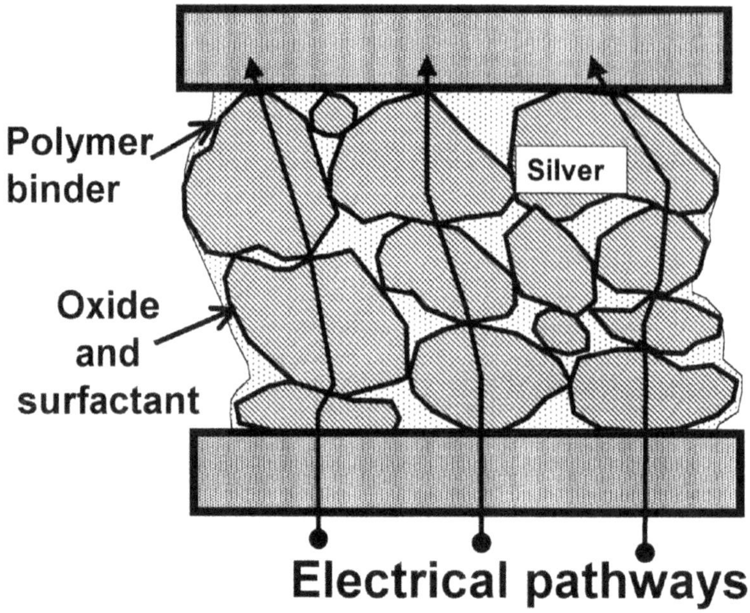

FIGURE 6.9 ICA model.

6.5 THE ASSEMBLY PROCESS

FC assembly processes are more numerous than for packaged SMT because of the various types of joining materials, the use of self-contained joining systems, and the need for underfill that can be applied at different stages in the process.[19]

6.5.1 Solder Reflow

Most FCs are joined to the PWB or package substrate by the solder reflow process. A number of solders are used, ranging from tin-lead eutectic to higher-melting allows made with a substantial lead content. Newer lead-free alloys are also available. The source of solder is typically the bump, but solder paste is also used, especially for high-melting and nonfusible bumps. Flux is applied when the solder bump is the source of solder, and its two main purposes are to hold the chip in place and to remove any oxides on the bump or substrate pads. Solder paste contains its own flux. The solder reflow process involves heating the assembly with the chip in place at a predetermined rate called a *profile* that is nonlinear but generally has a time/temperature profile resembling a steep bell curve. The first heating stage activates the flux to promote oxide reduction. The temperature is then increased, peaking at 20 to 40°C above the reflow point of the solder, allowing the bump to melt and form a metallurgical joint with the substrate metallization. The temperature is then reduced, permitting the solder to harden. The entire heating and cooling process is called *solder reflow.*

SMT Process with Eutectic Bumps. The eutectic solder reflow process is essentially identical to SMT assembly and closely resembles BGA attachment. The FC is an SMT device, the very first ever, and the modern FC is often a BGA. Eutectic solder bumps become the source of the joining material, and no additional metal needs to be added. Additional solder is sometimes provided as solder-coated pads or even solder-bumped substrate. The simplest approach is to use the bump as the joining material, however. Flux is usually required and can be added to the FC bumps or to the substrate. Many pick-and-place machines can be purchased with flux applicators, such as a rotating drum. This device provides a precise thickness of flux that can be applied to the bumps on an FC or other BGA simply by dipping the package into the material. Setting a doctor blade, positioned over the drum, to a specific gap controls flux thickness. The rotating drum therefore becomes a flux reservoir with just the right thickness of flux. Some rotating drums have channels or troughs of a specific depth that provide the reservoir. Several different depths can be used on the same drum to provide more than one thickness of flux. This feature is useful when different types of FCs, CSPs, or BGAs are being assembled onto the same PWB. The total amount of flux per bump is controlled primarily by the thickness of the reservoir, but withdrawal rate has a secondary affect. The drum rotation is synchronized with placement so that the drum is stopped when the chip is dipped. One disadvantage of the flux drum method is that the total assembly rate may be cut by up to 50 percent because of the extra step. A major advantage is that flux is controlled precisely and deposited only where needed, at the joint.

Flux also can be applied directly to the PWB by needle deposition, screen printing, or spraying. Many assembly machines can be fitted with needle dispensers for applying flux. Machines with needle dispensers only experience a minimal delay. Flux that is preapplied by a printer, high-speed automatic needle dispenser, or other means gives maximum throughput. Printing is a parallel process that can improve throughput significantly when PWBs are used with multiple bond sites, as with BGA manufacturing. Integrated lines have throughputs of over 10,000 units per hour.

Once flux is applied and the chip placed, the assembly moves through a reflow oven. A flux with chip-holding tackiness is preferred so that vibrations do not misalign the chips before solder reflow. Profiles are somewhat similar to those used for solder reflow, but fine-tuning will optimize results. The manufacturing goal is to place and reflow both FCs and SMDs on the same line concurrently.

Preapplied Flux/Underfill Assembly. Now that FC assembly is fairly well understood and reliable products are in the field, major efforts are underway to increase productivity. The flux and underfill processes are clearly the bottlenecks. Work has been underway for several years to combine these processes with some success. Preapplied flux/underfill is commercially available and can be used on standard FC and some SMT lines. The liquid flux/underfill is applied to the substrate before placing the chip. A stand-alone dispenser or screen printer can apply the materials, or the operation can be incorporated into the pick-and-place unit. Once again, note that dispensing by the pick-and-place equipment will drop productivity.

A needle dispenser can be used to apply a drop of flux/underfill onto the FC placement site. Work by Georgia Institute of Technology[20] and others has shown that the drop height should be high enough so that the FC forces liquid from the center to the perimeter of the chip. This helps displace air, a major concern for this process. Most success has been found with smaller dies, less than 10 × 10 mm, because of the increasing problem of displacing air with larger ones. Also, the preapplied underfills typically do not contain filler because it interferes with the soldering process. This means that the CTE of this class of underfill can be well above an optimal value, and reliability may not be adequate for large chips.

Once the chip has been placed, the assembly moves to a reflow oven. The chemical timing must be nearly perfect: Viscosity drops, flux activates, solder melts, joints form, and the polymer hardens. The rheology curve must be just right for good solder joint formation. Premature hardening of the polymer will inhibit good joint formation or even prevent attachment. The desire is for all the polymerization to occur during reflow, but a postcure can be required. In fact, some systems are designed to be reworkable until after a lengthy postcure, but this can be counterproductive. In some cases, the reflow profile may need to be extended to complete the cure. The preapplied materials and processes are still in a relatively early stage, and improvements can be expected.

Solder Paste Processes. Nonfusible and high-melting-alloy bump FCs generally are assembled by applying eutectic solder paste to the carrier or PWB. Both screen printing and stenciling have been used, but stenciling gives better resolution and control. Solders made with finer powder will give better results, especially for fine pitch. Vendors are providing paste specifically designed for FCs. Any paste that will be placed in production should be tested for compatibility with underfill if used. Most of the FC solder pastes, made with epoxy-based flux, are compatible with all underfills. Fine-solder dots also have been achieved by needle dispensing, but this process should be considered only for special circumstances.[21] Low-alpha-particle solder pastes are now becoming available because alpha emissions can interfere with IC operation.

6.5.2 Thermomechanical Attachment

Thermal compression bonding, long used for TAB, has been applied to small FCs on flexible substrate.[22] Gold bumps must be used with gold pads on the circuit. The chip is forced against the PWB under heating conditions. This is really a throwback to TAB, but without a "window." In fact, one of the earliest "no window" tape chip carriers was called *flip chip strip*. It may be noted that early TAB and some more modern versions did not have windows. The resulting gap between chip and substrate may be only 10 to 15 μm, and very fine filler underfill is needed, although some success has been achieved with unfilled polymers.

6.5.3 Adhesive Bonding

All types of electronic adhesives can be used here. Generally, adhesives provide the advantages of low-temperature bonding and wider compatibility. This can be crucial for certain optical FCs and temperature-limited substrates such as polyester film. Additionally, adhesives are compatible with thin-film circuits and indium–tin oxide on glass that cannot be soldered. All types of adhesives are also compatible with gold bumps that can only be soldered with very special alloys.

ICA Assembly. These paste materials are best applied by stencil printing, but screen printing, needle dispensing, and dipping are alternatives. The stencil should be no more than 100 mm thick. Laser-cut stencils will work, but electroformed stencils are considered best. The adhesive is printed onto the wiring board, and the chip is placed. The adhesive usually will be tacky enough to hold the chip in place as the assembly moves into an oven. A reflow oven can be used, but the temperature can be set much lower, typically 130 to 150°C. Many adhesives can be cured at reflow solder temperatures, and this would be an advantage where other components are being soldered. Nearly all types of bumps can be assembled with ICA, including eutectic Sn/Pb, but not all adhesives will form a highly stable junction. Virtually all are reliable with gold-stud bumps[23] and gold-plated nickel.

Once cured, the adhesive joint is fragile, more so than solder, and underfill is required. Standard underfills work well, and since no flux is used, compatibility is not an issue. The underfill will shrink slightly during curing, and this will tend to compress the adhesive joint and increase conductivity. The underfill also will greatly reduce any chance of silver migration. The underfill increases the thermomechanical reliability of the adhesive joints and greatly reduces the two most significant limitations of these solderless materials, mechanical strength and silver migration. Unlike all but the specially purified lead-based solders, silver ICAs do not emit alpha particles that are especially detrimental to memory and high-density CPUs. Adhesives are therefore a good choice for many FC applications.

ACA Assembly. ACAs have been used with FCs for a dozen or more years. In many ways, this is an ideal system. The ACAs are capable of attaching fine-pitch devices without concern for shorting, making them well suited for FCs. The ACAs require a very flat surface and do not tolerate noncoplanarity. Fortunately, FCs are quite planar, and most bumping gives narrow-range bump height tolerances. The ACA film can be preapplied to the chip or substrate by "tack bonding" with a short duration of heat and low pressure. The FC can then be placed onto the bond site. Both heat and pressure are needed, and the total force can be substantial. Dwell time can be a minute or longer, and some systems need a postcure. This means that ACA FC assembly is not really an SMT process, and assembly rates are slow. However, no underfill is needed, and this helps offset the time penalty. The ACA assembly can give reasonably good performance, especially on glass. Flat-panel displays use the ACA process to assemble FC drivers, and the concept is called ***chip-on-glass*** (COG).

Nonconductive Adhesive Assembly. The process, first reported by Matsushita, involves applying ultraviolet-cured underfill to the bond site, pressing an FC against a glass flat-panel display circuit, and curing the adhesive by exposure through the glass from the opposite side. The adhesive is designed to shrink and create a holding force that translates to pressure at the bump-pad interface. The process, although limited, has worked well for COG applications.

6.5.4 Testing and Rework

Several testing and rework methods are available. IBM developed the first one many years ago as a method for producing known good dies (KGDs). The chip is soldered to a reusable ceramic test substrate that has undersized bond pads. The chip can then be tested and burned in. The chip can be removed by breaking the purposely weak bonds. The bumps are then reflowed to regain their original spherical shape. However, most FC assemblies also can be tested and reworked provided that there is no underfilling. A bad component can be removed by applying localized heat to melt the solder joints and then pulling off the chip. Excess solder can be removed from the bond site with a heated copper block. Once tested "good," the assembly is underfilled. A few of the newer underfills are said to be reworkable, but this involves tedious steps and mechanical cleaning of the site. Test and rework therefore should be done before underfilling.

Pretesting. Pretesting of FCs before assembly probably cannot be considered by the assembler. The section, "Known Good Die Challenge," below describes available methods, however.

In-Circuit Testing. FCs can be assembled by soldering or adhesive bonding and then tested electrically prior to underfill. At this stage, rework is possible. Early FC technology used ceramic substrate with no underfill, making rework relatively easy. The modern FC on organic substrate requires underfill that generally is not reworkable, although considerable work is underway to address reworkability. The best production strategy when FC removal will be required is to assemble the FC by solder reflow, test, and rework as necessary and then underfill. Those who want to test after underfilling will need to wait for advances in underfill that will be described in the next section.

What about adhesives? Can they be reworked? Most ACAs are not reworkable in the practical sense. The thermoplastic class can be reworked, but this requires applying heat to debond and removing adhesive that can stay on the chip, substrate, or both. ICAs require underfill and are even less reworkable than solder. The best answer to a reworkable FC appears to be the FCIP (flip-chip-in-package). The FC is connected to a BGA substrate that can be soldered to the PWB and removed using BGA rework techniques. FC BGAs are becoming more prevalent and are available in CSP format and larger sizes.

6.6 ENCAPSULATION/UNDERFILL

Silicon, the common substrate for semiconductors, has a very low CTE. All other substrates, including ceramic, have higher CTEs. The popular organic PWBs have CTE values that are about an order of magnitude higher than that of FCs. The large thermal mismatch produces significant stress during thermal cycling. Heating causes the board to expand to a greater extent than the chip, whereas cooling produces the opposite result. The net effect is that the joints are strained, and fracture failures occur. Although it is possible to produce laminates with lower expansion rates, most are matched to the CTE value of their copper conductors, 17 to 18 ppm/°C. No practical joint material can survive, not solders and not conductive adhesives. The result is that low-cost FCs on organic boards cannot pass most reliability criteria. Fortunately, there is a simple solution called *underfill.* Figure 6.10 shows the thermal-mismatch problem.

Underfill can be viewed as a laminating adhesive that mechanically couples the low-expansion chip to the high-expansion board. PWB movement is constrained by the very high modulus chip. Since the board area under the chip moves approximately in step with the FC, stress is reduced, and the joint is preserved. The chip, joints, and PWB are locked in step, and the thermomechanical stress is moved away from the critical assembly zone into the laminate that can tolerate the forces. The underfill must have very special properties to accomplish this feat, including the correct modulus and CTE. A study of Fig. 6.11 will show that the CTE should match the properties of the joint, not the

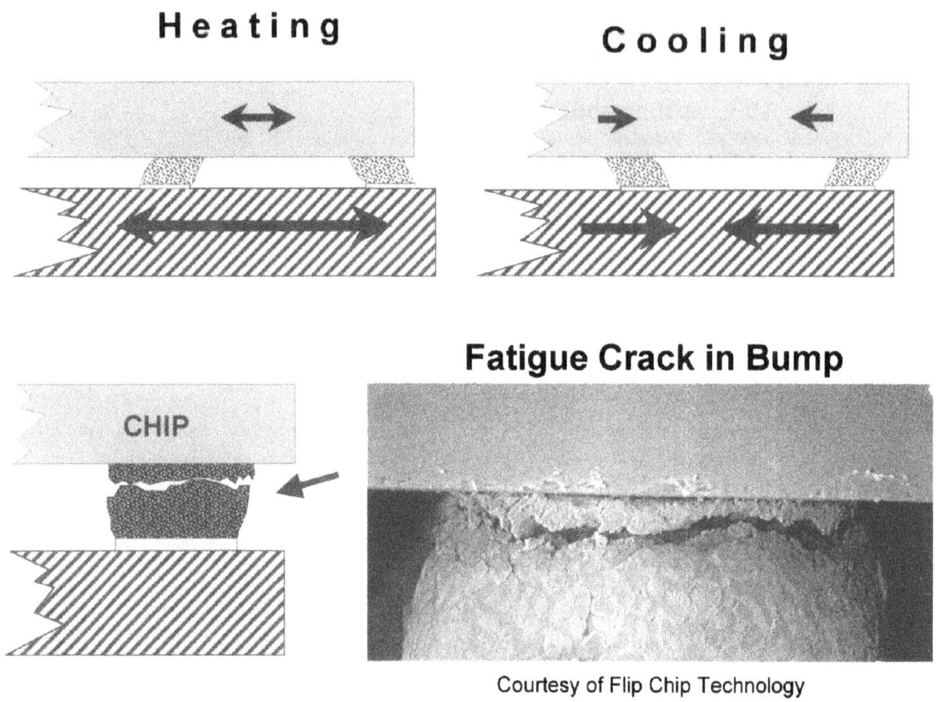

FIGURE 6.10 Thermomechanical mismatch model.

chip. If the underfill has a very high CTE, heating will elongate the joint, whereas too low a value will cause compression. Either way, fatigue will be induced in the vertical axis by the underfill, even though lateral movement will be controlled. The next subsection will describe all possible types of underfill, including those not yet invented.

6.6.1 Preapplied Flux/Underfills

This process was discussed earlier. It will suffice to mention here that the preapplied flux/underfill system eliminates underfill as a separate step. However, if a postcure is required, then only part of the underfilling process is eliminated.

Liquid on Substrate. There are several possibilities for a preapplied underfill, but the simplest is a liquid that can be deposited onto the PWB just prior to chip placement. The material can be applied by a conventional stand-alone needle dispensing machine or by a dispenser that is part of the FC placement equipment. Screen or stencil printing also can be considered if strips of substrate or large arrays are being assembled. The preapplied-on-substrate materials have been referred to as *no flows*, a handy but inaccurate term.

The process involves depositing a specific amount of flux/underfill, placing the FC carefully into position, and running the assembly through a reflow oven. There is a tendency to trap air as the chip is pushed into the liquid, and considerable work has gone into the process, especially at the Georgia Institute of Technology.[24] The best method for reducing trapped air is to deposit a glob of material onto the PWB bond site and force the chip down so that material is displaced from the center outward. Good results have been achieved with smaller dies (<10 mm). Even if better techniques will allow larger dies to be placed with few voids, there may still be a maximum die size limit related to the cured properties. The liquid preapplied products have a high CTE (60 to 80 ppm/°C) because

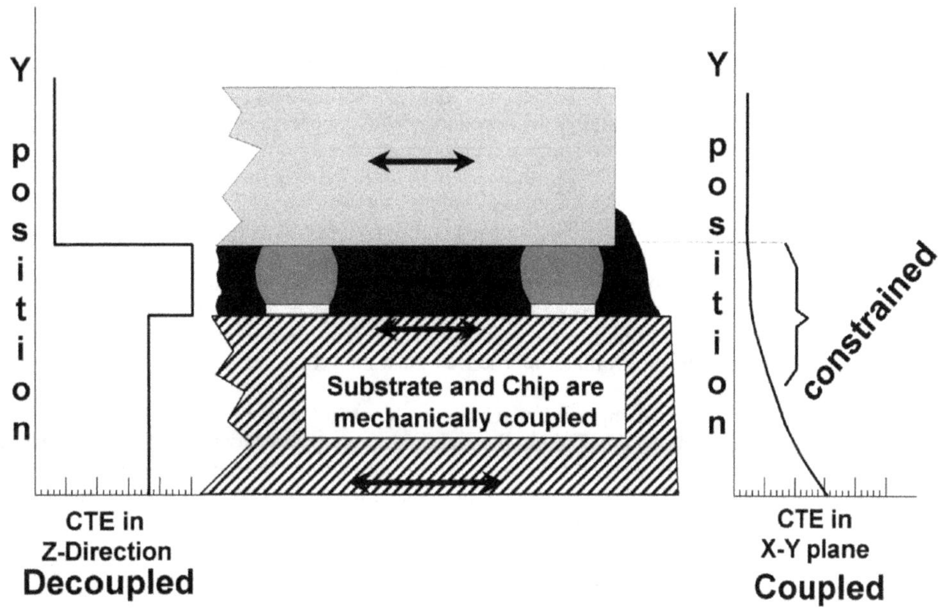

FIGURE 6.11 Underfill mechanism.

expansion-reducing filler should not be used, which would interfere with solder joint formation. Larger dies produce larger stresses, and the unfilled flux/underfill may experience problems. Nevertheless, preapplied liquid underfills enable in-line FC processing with a major boost in productivity.

Solid on Substrate. You may be surprised to find that underfills can start off as solids. Solid materials can be applied to the substrate as a film or as a liquid that is hardened before chip assembly. Solid flux/underfill on the substrate could allow wiring boards or chip carriers to be handled and even shipped in ready-to-bond form with all the materials required for assembly. The flux/underfill would need to go through a liquid stage during assembly, of course, and then return to a solid state again. The final solid state could be higher melting or even nonmelting (thermoset) depending on the polymer system. For example, low-melting, B-staged thermoset film could be used. The material would melt during solder reflow but then polymerize to a high-performance cured underfill. A thermoplastic could even be used, and this would result in a reworkable underfill. However, relatively high temperature and pressure on the FC probably would be required. A product and the necessary bonding equipment have been made available by Nitto-Denko.[25] The idea of solid underfill is analogous to ACAs, except that there is no conductor in the polymer film. While the concept of solid flux/underfill on substrate may have value, the process does not fit the infrastructure because a special bonder is required, and this may be a significant limitation.

Solid on Chip/Wafer. One also can apply solid flux/underfill to a chip or wafer.[26–28] The ramifications are profound. The flux/underfill application is moved to the semiconductor industry and away from the assembler. The resulting FC becomes a ready-to-bond component. The underfilling process appears to go away and becomes transparent to the assembler. The FC is now a CSP. Most important, the FC with integral flux/underfill fits the SMT process; FC = SMT! And if this is not enough, solid flux/underfill on chip can be made reworkable.

Work has been in progress since the mid-1990s to develop wafer-level flux/underfill.[29] Several segments of the industry, as well as consortia, are involved, and the U.S. government has contributed financial support. Some of the players are LSI Logic, Motorola, Loctite, and National Starch and Chemical.[30] Cookson Electronics, with corporate partners, including Alpha Metals and Speedline, is funding a consortium with several universities, including Georgia Tech and Binghamton University, as prime contractors. Semitech is also sponsoring work. The substantial effort by the electronics industry almost certainly will produce results. Success can shift micropackaging preference toward FCs. Others have investigated multilayer combinations of materials on-chip and on-board, although the increased complexity would seem to unnecessarily limit these systems.[31]

Several strategies are being explored for wafer-level FCs, but they can be divided into single material and multiple. The simplest multilayer material consists of separate underfill and flux layers. While the method of application is more involved, material development is straightforward. The underfill can be a filled thermoplastic (reworkable) or a thermoset. No flux activity is required, and filler can be used. The flux layer can be applied over the solid underfill layer or just on the bumps. Coatable solid flux is already available. The flux layer contains no filler because it would interfere with solder joint formation. The flux must melt, bond, deactivate, and convert to a higher-melting solid.

The single-component material must have flux properties initially but then convert to an underfill. This can be done within the realm of epoxy-based flux because flux agents such as carboxylic acids also can serve as hardeners. Work continues in this area, and a product may even be commercial by the time this book is in print. It may even be possible to produce a linear thermoplastic epoxy that will be reworkable.

6.6.2 Postapplied Materials

The term *postapplied* is used to describe the common capillary flow underfills that are dispensed under the FC after it has been assembled to substrate. This is the more common approach, and it can be viewed as the standard process for now.

Capillary Underfill. The common capillary flow underfill is a postapplied product. The FC is first assembled by soldering, conductive adhesive bonding, or other means. Then the underfill is allowed to flow under the chip. The underfill is "pulled" along by capillary action or intramolecular attraction between the liquid and the solid surfaces. Flow requires a set of criteria that include an "advancing" wetting angle that results when the surface tension of the underfill is less than that of the solid surfaces. The viscosity must be low enough to permit flow within a reasonable time, although heating can be used to reduce viscosity and increase flow rate. Filler, added to reduce the CTE, must be small enough so that particles do not block flow. A good approximation is that maximum particle size should not exceed one-third the gap size. State-of-the art materials offer fast flow and fast cure.[32–35]

The capillary underfill must be applied using a pattern that does not trap air. A general rule is to apply underfill from only one side of the die for larger (>10 mm) FCs and then dispense more material on the other three sides after the underfill has flowed to these sides and displaced the air. Smaller dies can use an L-shaped dispensing pattern. Material is dispensed starting at one corner of the chip, and the dispensing needle is moved the full length and then around the corners, as shown in Fig. 6.12. Once the underfill has flowed completely under the chip, adding the missing L can complete the fillet. It is not always necessary to add a fillet, but reliability is maximized. A faster process only requires that a dot of underfill be added to the far-side corner of the chip after flow out, as shown in Fig. 6.13. Reliability is slightly lower, but productivity is higher. The L plus dot approach works well for smaller chips used in cellular phones and pagers. The dispensing equipment industry has reported extensively on the underfilling process.[36–38]

Once underfill is cured, rework is not possible when using any of the commercial capillary flow products. However, work continues in the capillary underfill area to produce reworkable materials.[39]

Encapsulant (Over Chip). Both liquid encapsulant and solid molding compounds may be used to protect FCs used in packages. Underfill must be applied first so that there is no airspace underneath when the encapsulant is applied. Many FC BGAs have been produced with no overcoat or encapsulant

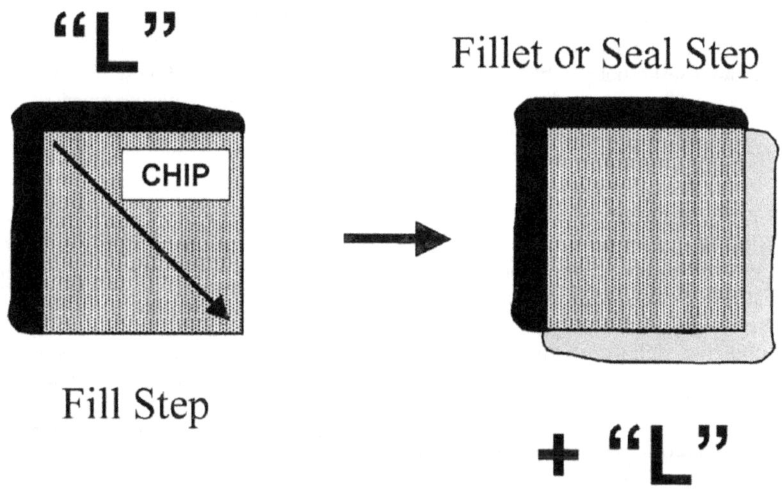

FIGURE 6.12 Double-L dispense pattern.

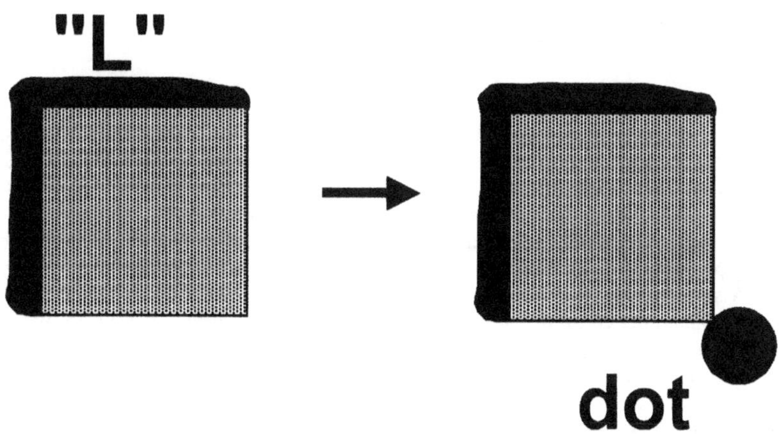

FIGURE 6.13 L and dot pattern.

because the underfill protects the sensitive underside of the die. Also, the backside of the die can be used to radiate heat by applying a heat sink. Once encapsulant covers the die, the heat-sink option is eliminated. Work is also underway to transfer mold encapsulant over the chip while simultaneously underfilling.

6.7 SUBSTRATES FOR FCs

The first FC substrates were ceramic, and they are still preferred from the viewpoint of performance and ease of assembly. Flatness and low CTE are two preferred attributes for FC substrates that are inherent in most ceramics. During the 1990s, organic substrates became popular for many FCs, especially the smaller, lower-lead-count variety. However, the move to organic substrate introduced the problem of significant thermomechanical mismatch because of the relatively high CTE of organic materials. However, underfilling adequately reduces joint stress, and reliability can be regained. Underfilling, while allowing popular, low-cost substrates to be used, increases processing time and adds cost. Nonetheless, underfill allows virtually all substrates to be used. The preferred organic substrates include standard FR4 circuits, higher grades of epoxy such as bismaleimide triazine (BT), and flexible circuit substrates. Large, high-performance chips, such as CPUs, continue to favor ceramic materials.

6.7.1 Ceramic

IBM first used ceramic circuits and chip carriers for FCs. Ceramic offers two important advantages for FCs. First and most important, the CTE is low enough that there is only a small thermomechanical mismatch between chip and substrate. This means that underfill is not required for thermomechanical enhancement, at least for smaller dies. The second feature of ceramic is high-temperature stability as well as superior thermal conductivity. Ceramic can handle 300°C or higher, a valuable feature for a chip carrier. Some materials conduct heat better than aluminum and therefore are ideal for heat dissipation.

Although ceramic has become less popular in recent years because of higher cost, it has been used to build FC BGAs. AMD, for example, has used ceramic BGAs for the K6 and K7 microprocessors. The high-temperature performance of ceramic allows higher-melting solders to be used for chip attach so that melting will not occur when the package is soldered to the PWB. Ceramic substrate is also much less prone to warping than organic laminate and is a good material for FC BGAs. The factor that will continue to reduce usage is cost, especially for high density. The PWB industry continues to advance methods, such as microvias, that improve the cost/density picture, and this will likely increase the cost differential.

6.7.2 Organic, Rigid

Organic laminate has become the preferred substrate for both FCOB (flip-chip-on-board) and FCIP (flip-chip-in-package). In fact, second-generation FC is really centered on organic substrate to lower cost. The advent of the BGA in the early 1990s, especially the plastic BGA (PBGA), pushed the demand for better laminates. New materials were introduced, and more are under evaluation aimed at providing improved mechanical stability and higher heat ranges. BT resin has become one of the most popular advanced laminates and is now used widely for BGAs. The same high-performance BGA substrates also provide FCs with desirable properties. Smaller chips used on cellular phones, pagers, and other consumer products do not necessarily require higher performance than offered by FR4. It is the FCIP products that need better substrates, and these have been developed for wire-bonded BGAs.

6.7.3 Organic, Flexible, High-Temperature

Flexible circuit materials have long been used for chip carriers. TAB, called *tape carrier package* (TCP) by Intel, has been used to handle high-density chips with very good results. The polyimide class of substrate, such as Dupont's Kapton, is noted for extraordinary high-temperature performance. Some of these materials have a zero strength factor at 800°C, a region usually confined to ceramics. The flex materials are also thin and compliant, making them well suited for use as chip carriers and FC assembly substrates. A number of companies, including IBM, 3M, and Amkor, have developed products that can be classed as *flex-based packages.* Both wire-bonded and FC assembles have been made. Thermal compression bonding to flex is also possible, and Motorola has developed technology here.

Flexible circuitry is an enabling technology for many industries such as disk drives. The flex allows the read/write head to move over the disk.[40,41] Faster and higher-density drives have been made possible by using lighter components placed closer to the head. Today, the FC is the optimal solution for high-density magnetoresistive disk drive technology. The FC is assembled to the polyimide drive circuit using reflow soldering. Stud bumps of various materials also have been used for flex.[42] Underfill must be used even though the thin, compliant flex produces lower thermomechanical stress than hard board. Although it is possible to make flex substrate with a low CTE approaching that of the chip, a value of around 18 ppm/°C is used in order to match the copper conductors. This means that underfill is required.

Disk drive makers, most notably Seagate, have been able to apply underfill without a fillet step. Underfill is applied from one side, and the assembly moves to the curing oven. Underfill with lower filler and hence lower viscosity can be used, and this produces self-filleting. The high-volume, low-cost requirements of the disk drive industry suggest that this segment of electronics will continue to lead productivity and cost reduction. Expect to see the implementation of preapplied flux/underfill here.

6.7.4 Organic, Flexible, Temperature-Limited

There is a large segment of the flexible circuit industry that uses very low cost polyester substrate such as Dupont's Mylar. The polyesters have a temperature-processing limit of about 150°C, so one probably can rule out solder assembly even though low-temperature alloys exist. This is the area where conductive adhesives excel. Both aniso- and isotropic types can be used. Two basic types of conductors are used with polyester film, and they are traditional subtractive copper and additive polymer thick film (PTF). Bare copper can be used with adhesives but not without junction stability concerns. A much better interface is gold plate over nickel plating on the copper. All adhesives form a stable electrical junction with gold-plated conductors.

PTF silver ink makes a very good interface with conductive adhesives, as would be expected because of the similarities. PTF ink has a silver-rich surface that mates well with the silver matrix of ICAs. Millions of such junctions have been made without any issues using SMDs. ACAs also form good connections with PTF ink because the conductive spheres tend to embed into the ink surface. The oxide that forms on silver is of no real consequence because it is electrically conductive. Radiofrequency identification (RFID) products are now being made with PTF inks and FCs bonded with adhesives, and they will be covered later under "Applications."

6.8 FEATURES AND BENEFITS

FCs offer a wide variety of benefits that often seem to be at different ends of the property spectrum. One company may use FCs to reduce cost, another for high performance, and a third to reduce size. Let's examine the features and see what benefits can be derived.

6.8.1 Geometric Considerations

The FC is used for two primary reasons: maximum performance and minimum geometry. The FC represents the simplest, smallest, and lightest minimal package that is possible. No other systems can be smaller, have a lower profile, or be lighter. This is so because the FC adds only the minimum essentials to the bare chip. The FC can be viewed as a chip with a simple connection interface for direct attachment to substrate. The bumps can have very low height so that the resulting assembly has a height only slightly greater than the thickness of the die. The footprint is exactly equal to the dimensions of the die, although some allowance may be needed for the underfill fillet that surrounds the chip. In the future, however, emerging wafer-level underfills may reduce the fillet or even eliminate it. Thus, when the very smallest size, lowest profile, or minimum weight is essential, FC is the best choice.

Footprint. There is nothing smaller than the bare die, and an FC structure adds nothing at all to size. No CSPs can be smaller than the FC. One could argue that underfill increases the footprint because of the fillet and a "keep out" area that must be reserved for excessive flow out. However, even this minor addition of area is about to change with wafer-level underfill. The wafer-level flux/underfill solids that will be an integral part of the chip are not expected to flow out to any significant degree. Applications that require minimum area can take advantage of the FC. The FC can be used for very high I/O chips; not so with most CSP technology.

Profile. Even the FC connection has the lowest profile possible. Joints as small as 10 to 15 mm are being used. Even the more typical 50-mm high joint does not add significantly to the height. Many of the CSPs, while keeping area close to chip size, can more than double the height of the final assembly. FCs have been used in handheld products where reducing thickness was more important than area. The StarTac cellular phone was said to use FC technology for height, not footprint reduction. A second, "stacked" wiring board could then be added, thereby doubling the useful board area. Smart cards and RFID products are also benefiting from the minimal profile. FCs should be considered when profile reduction is a key consideration.

Weight. Once again, nothing can be lighter than the basic die. The interconnecting structure is as minimal as possible, consisting only of the short bump that forms part of the entire bump. The bumps are often so small, less than 50 mm in diameter, that their weight is insignificant even when solder is used. However, lower-density conductive adhesive reduces weight even more, and underfill has a low density of around 3 g/cc depending on filler loading. FCs should be considered when weight reduction is an important factor.

6.8.2 Performance

Speed. Many packages limit the operational rate or clock speed of a device. Added wire length adds signal delay time. Many of the CSPs have a die-sized foot print, but their wiring structure adds length and signal delay compared with an FC. It should be mentioned that an IC designed for FC packaging will optimize the on-chip wiring so that rerouting is not required. The move to copper IC conductors also will reduce internal resistance, making in-chip routing more effective. Chips designed for perimeter pads and then rerouted externally can have a very inefficient layout that just about doubles the total wire length. Short lengths reduce performance-robbing inductance. No package has lower inductance than the simple FC bump.

I/O Density. FC is the clear winner for high density. FCs are in production with more than 2000 bumps. IBM has built working prototypes with 10,000. Advanced bumping technologies, such as solder plating,[43] allow extremely fine bump pitch. It is safe to say that the FC will never be the limit to density; it will be the substrate.

Process-Related Economics. FC has been designed as a wafer-scale process. Most bumping methods deal with the wafer and therefore enjoy considerable economics of scale. The various parallel or mass bumping processes therefore should achieve the same scaling economies as the IC. The move to 300-mm wafers will reduce bumping costs.

Underfilling is still a board-level serial process that reduces the overall economics and productivity of the FC. However, this is about to change as wafer-level flux/underfill developments reach fruition. In the future, all the materials needed for FC assembly will be applied at wafer level. Instead of applying underfill to each assembled FC, the entire wafer will be coated. Thousands of FCs will have underfill applied in minutes, not the many hours now required for the capillary flow systems. In the meantime, preapplied flux/underfill is helping to reduce the processing time and boost the productivity. However, the wafer-level integral flux/underfill FC of the future likely will provide the best economics of any micropackage.

6.9 LIMITATIONS AND ISSUES

First, let's realize that traditional packages will continue far into the foreseeable future. SMDs and even feed-through devices have a continuing role to play, and the infrastructure for their manufacturing and assembly is well established and entrenched. Therefore, let's look at FC limitations and issues compared with other micropackages.

6.9.1 Known Good Die Challenge

Can FC deliver known good product? The answer is yes and no. Wafer-level testing continues to advance, but most people feel that it is an approximation at best. Recent advances in spring probes, notably by Form Factor, may be changing this, but for now, we cannot totally rely on wafer-level testing. Several schemes have been applied to FC test and burn-in over the years that can ensure good chips, but at a cost. IBM has assembled FCs for test and burn-in on ceramic substrate using a restricted pad technique. The pad on the test module is smaller than would be used normally for the specific chip. The assembly, which can contain many chips, is run through a test/burn-in sequence. The FCs are then removed by mechanically fracturing the minimal joints.

NCMC developed substrate and a process for test and burn-in that does not require mechanical fracturing.[44] A thin-film ceramic test circuit is produced that will accommodate FCs and allow attachment by solder reflow. The assemblies are burned in. Next, the chips are removed by melting the solder joints and removing the FCs. The special pads allow complete removal with all the joint solder remaining with the FC because the special test circuit metallization produces a dewetting effect. The test substrate can be used many times before the pads need to be renewed. This process, as well as electroplating bumping services, has been commercialized by Unitive, Inc., Research Triangle Park, NC.

Various probing technologies are also available, including IBM's dendritic contact pads and Form Factor's microsprings. One can still argue that these methods are not as useful as on-site testing and burn-in afforded by CSPs, however. Yet the intensifying effort to resolve these kinds of issues is likely to produce results in the near future.

6.9.2 High-Density-Circuit Requirements

The packaging revolution of the 1990s was the result of continuing advances in the semiconductor industry as well as the consumer demand for higher performance and real portability at an attractive price. This can be summed up in the marching theme, "smaller, faster, cheaper." The perimeter package had run out of steam for the CPU, and the transition to area array was more than a nice idea—it was an absolute necessity. These major changes in packaging put significant pressure on the printed wiring industry. New materials and processes were needed both for chip carrier substrate and PWBs that would accommodate the area array and micropackages in general. Tried and true drilling is being replaced by laser machining to produce microvias. Effective use of microvia technology requires changing the old stack-up lamination method of making multilayers. Various builtup circuit conductor methods have moved into production, but only a few are able to produce the required high-density products, and they are mostly in Japan. The PWB industry eventually will add the new infrastructure required for this general area called *high-density interconnect* (HDI).

6.9.3 Assembly Difficulty

Smaller package features typically raise the bar of difficulty. However, area array products such as FCs are much better at self-alignment. FCs purposely misaligned by 50 percent are able to center during reflow. Work by Siemens indicates that a 30 percent misalignment will not reduce production yield. Many types of FCs also solve the predicament of applying solder paste to small areas. No paste is required because the bump contains the joining material. In all, the difficulty added by the FC's small size appears to be offset by the features just mentioned.

6.9.4 Ramifications of Die Shrink

The semiconductor industry achieves the objectives of smaller, faster, cheaper primarily by die shrink. The die is scaled down to a smaller size, undergoing a "shrink," to boost the speed through shorter connection paths while reducing cost because more dies can be placed on each wafer. It is possible to lay out the bump pads on an FC favoring the center area to accommodate die shrink without altering pad placement, and this is presently being done. Progressive die shrinks, often accompanied by minor IC architecture changes, inevitably require a new bump pad configuration that will not fit the original PWB pad layout. The designer is then faced with only a few choices for FCOB: Modify the PWB or reroute the FC —on-chip. The die shrink problem is reduced if FCIP is used because the chip carrier substrate can be rerouted and keep the same BGA footprint. The same kind of rerouting is also required for all of today's CSPs. Some bumping providers also offer rerouting on the FC.

6.10 PERFORMANCE AND RELIABILITY

FC holds the enviable title of logging the highest reliability of any system. IBM has established a long record of near-perfect reliability for FCs on ceramic. Insiders talk of thermocycling reaching the 60,000 mark before the test was terminated. The incredible performance on ceramic should not be a surprise, however. First, the ceramic circuits were designed to approximate the properties of the FCs with the CTE being only slightly higher. Smaller dies were used in these earlier mainframe modules, further reducing any stress. More fatigue-resistant high-lead solder was used. These MCM modules were cooled with highly efficient dynamic liquid heat exchangers. Many of the early modules are still running today.

Second-generation FC, the move to organic substrate, is not as reliable. It also should be noted that ceramic is still used today for both FCIP and FCOB. There are several factors that reduce the time to failure for modern lower-cost systems. The most obvious is the greater thermomechanical mismatch between chip and substrate. The substrate is also more prone to warping, and this can produce high strain forces. Solder joints are commonly made with eutectic tin-lead solder that work hardens and fatigues faster. The use of underfill to counteract the high mismatch is a double-edged sword. While correctly applied underfill can extend life well past 1000 cycles, voids and other defects can cause failures within the pass criteria. Small, undetected voids may reduce thermal cycle life by up to half, producing marginal reliability. Larger voids are detected more easily and are less likely to occur. Overall, FC performance is quite acceptable and steadily improving. Wafer-level flux/underfill may solve the problem of voids altogether and allow more consistent materials to be employed in the future.

6.11 APPLICATIONS

FC is somewhat remarkable in that it is used for the highest-performance, most expensive products, such as supercomputers, and the lowest-cost systems, such as smart cards. FC simplicity appeals to high-volume, low-cost products like the smart card and the RFID tag because of low cost and ease of assembly, where chips can be pulled directly from bumped wafers. The simplicity of having only a bump array structure with no external packaging provides the next desirable attribute of minimum size. The simple short-length bump and the resulting direct attachment structure also deliver the high electrical performance that is so appealing to makers of high-speed, high-end products. One added benefit of the tiny bump is that the highest density can be achieved so necessary for modern computer chips and other high-I/O products. The result is that the FC is used virtually everywhere, from lowest to highest range. All the major electronic business segments, including automotive, computers, and communications, have embraced FC. This is why many people believe that the FC is the final destination for future packaging.

6.11.1 Computers and Peripherals

Virtually all computer CPUs now require and use FCs. We have gone beyond the point where anything else can accommodate the large number of I/Os and high clock rates. Early conversions included the PowerPC from IBM/Motorola and AMD's K-6. In 1998, Intel's Pentium II went to FC. All are FCIP because this format allows test/burn-in and rework. The early FCIPs were ceramic, and some still are, but the Pentium moved to organic, and this will be the trend. The Pentium II FC has 2100 bumps, although some are redundant. This produces better heat transfer and a symmetry that allows better assembly and underfilling.

Computer components and peripherals, including disk drives, displays, and printers, have been quickly adopting FCs. Seagate went from SMT to COB (chip-on-board) to FC during the 1990s to increase performance and reduce cost. The read/write amplifier head is placed directly on the flexible circuit using solder reflow followed by underfilling. Even though flex is compliant and produces smaller thermomechanical forces than PWBs, reliability requires underfilling. However, the underfill can be applied without a secondary and time-consuming filleting step. The assembly goes directly into a curing oven once underfill is applied to one side. Even more productivity will be achieved when pre-applied flux/underfill is adopted in the near future. Figure 6.14 shows an FC on flex disk drive assembly.

6.11.2 Automotive

The automotive industry followed right behind mainframe computers to become a veteran area for FCs starting in 1968. Delco (now Delphi) pioneered this area and developed novel bumping and assembly techniques. As with early computer applications, the substrate was ceramic and still is, although the industry is just now moving to organic for such items as ABS. Figure 6.15 shows an engine controller with FC on ceramic. Automotive requirements are probably the most difficult within the industry.[45]

6.11.3 Consumer Products

There are now so many consumer products using FCs that it is a task just to keep up. The digital watch is certainly worthy of note and remains the highest-volume application. Figure 6.16 shows an

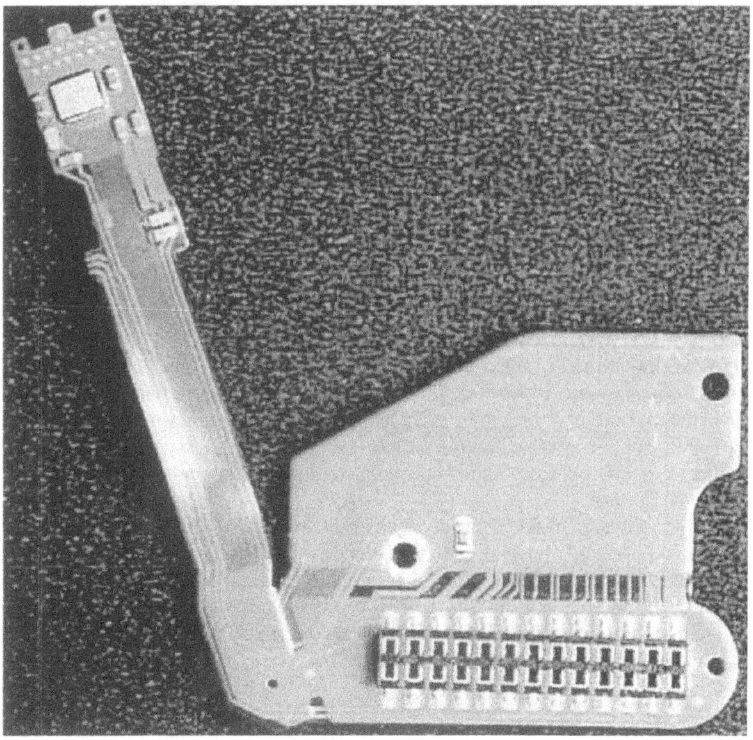

FIGURE 6.14 FC disk drive assembly.

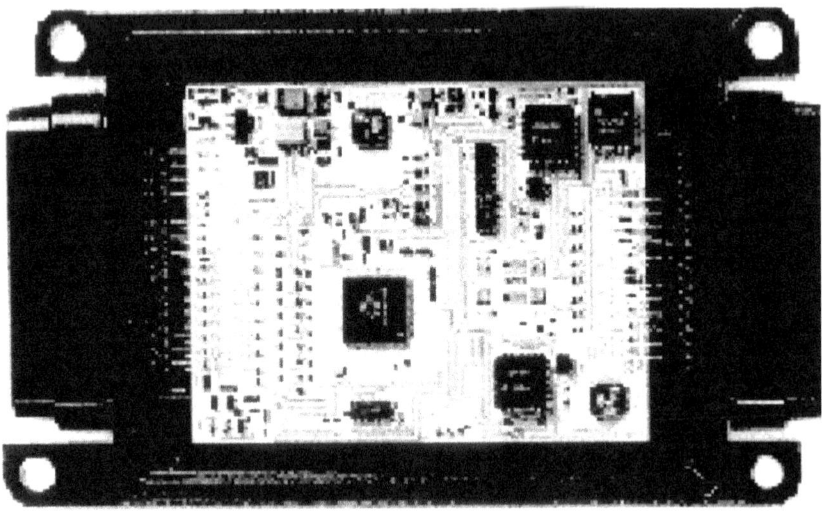

Delphi-Delco

FIGURE 6.15 Engine controller.

FC watch product. Perhaps the most famous FC example is the earlier Sony HandiCam digital camera shown in Fig. 6.17. Many of the FCs are CSPs because they are FCIPs.

6.11.4 Communications

Communications continues to be one of the most exciting, fast-paced areas of electronics development. Personal communication has boosted this market to one of the most important within our industry. At the same time, convergence and incredible miniaturization require adoption of the latest packaging technology. Motorola pioneered FC in this area beginning with the Pen Pager of the mid-1990s, shown in Fig. 6.18. The StarTac followed close behind and was launched as the world's smallest and lightest (99 g with battery) in the mid-1990s. Since then, many other phones have moved to FC, not just for size and weight reduction, but also for performance. Figure 6.19 shows the famous StarTac.

6.11.5 Smart Cards/RFIDs

We have seen that the FC can accommodate thousands of I/Os and let clocks run past 1 GHz. What about the opposite end—low cost, low I/O? The same question was asked about TAB many years ago when it was the premium high-I/O package. While the Pentium portables were using TAB in the late 1990s to get performance within a low-profile package, low-cost Japanese calculators used TAB to reduce cost. FC provides the same paradox for the low-end market—the highest performance delivers lowest cost.

The emerging area of RFID tags is enabled by FC. The RFID is essentially an antenna and an IC. The antenna picks up a burst of electromagnetic energy to power the chip and a signal transmitted by the reader. The IC processes information and then returns an "answer" by transmitting an rf signal through the antenna back to the reader. The simple circuit of a loop antenna and chip is built cost-effectively using DCA. The antenna is built on low-cost flex film, and chip connections are made with low-temperature conductive adhesives.[46,47] Figure 6.20 shows an RFID device.

6.11.6 Other FC Products

Although the military is often cautious and slow to adopt newer technologies, FC work has been done under U.S. government sponsorship.[48] A consortium of military and aerospace companies has

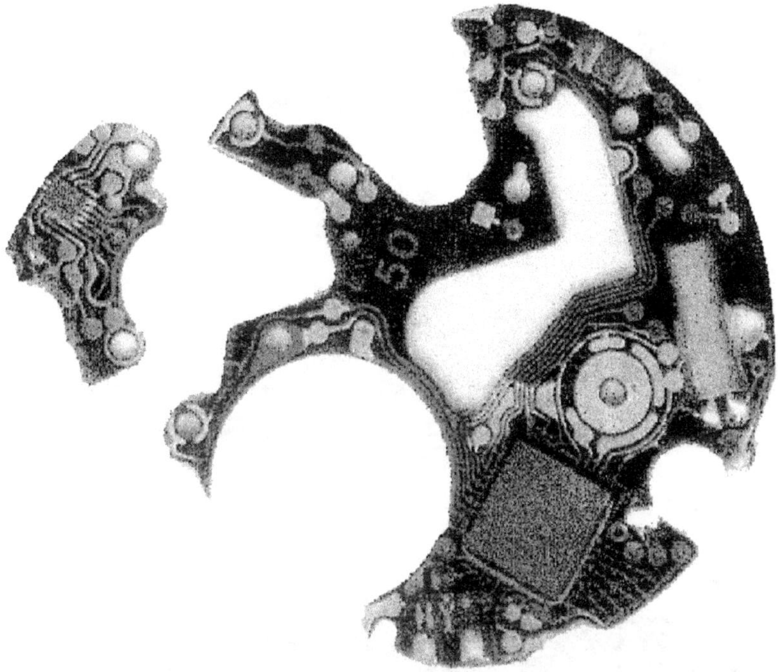

FIGURE 6.16 Watch.

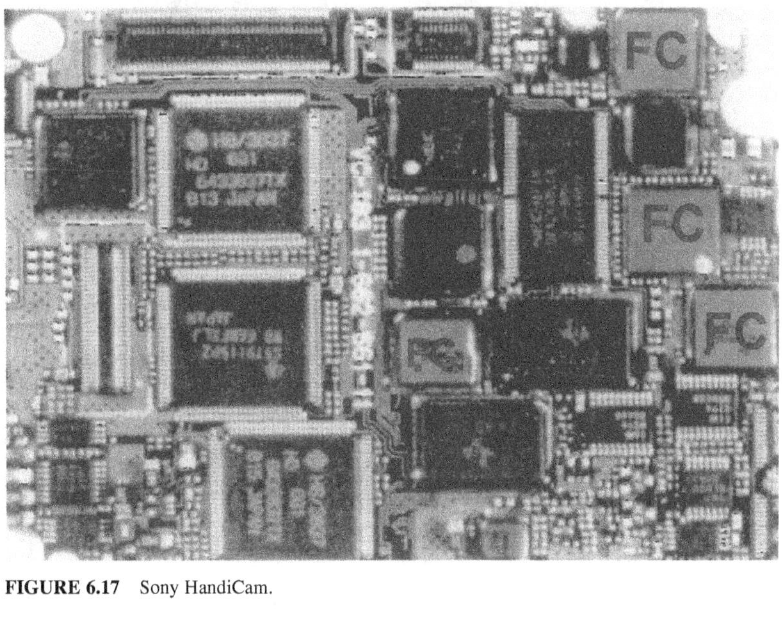

FIGURE 6.17 Sony HandiCam.

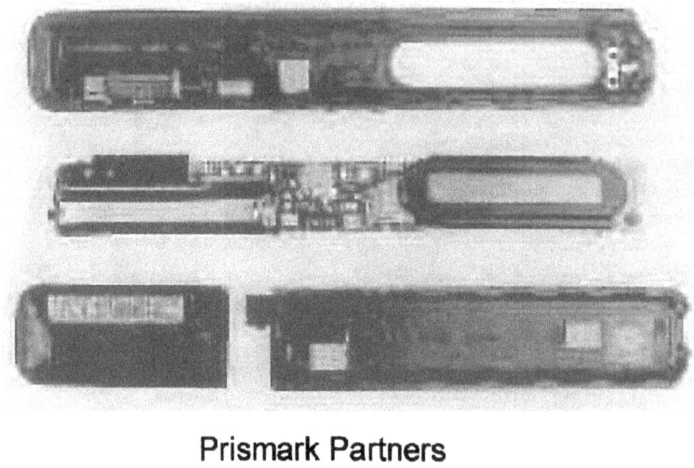

FIGURE 6.18 Motorola Pen Pager.

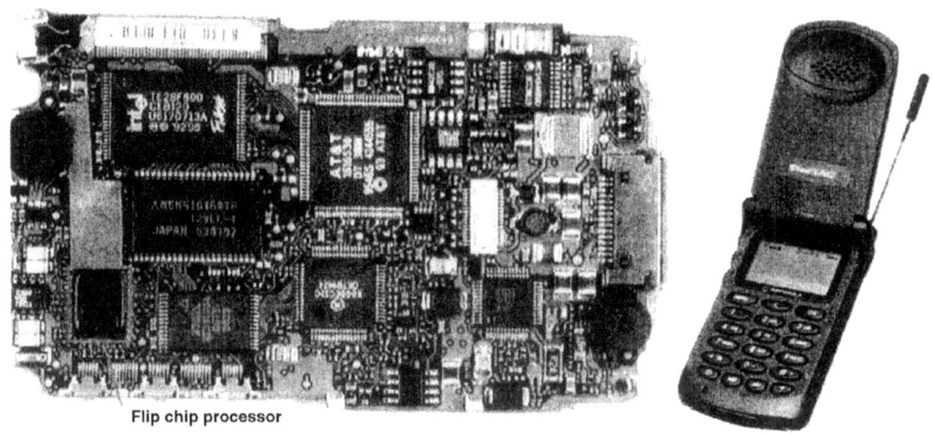

FIGURE 6.19 StarTac. (*Motorola.*)

done extensive development and evaluation of numerous FC technologies. Testing of FC on a missile demonstration module made of polyimide led to the conclusion that this area of technology can provide weight and size reduction in a durable and robust system.

The medical products field is also embracing FC technology, but perhaps at a more cautious pace common for that industry. Implants represent an ideal area where size reduction is so critical and the stable environment of the body eliminates thermomechanical stress. One fascinating product using FC is the DNA analyzer, a disposable detector that consists of a chip on flex.[49] A photograph of the detector module is shown in Fig. 6.21.

6.12 SUMMARY AND CONCLUSIONS

Many of us have long referred to FC as the ultimate package and the final destination[26] for micropackaging. Nothing can be smaller, run faster, or hit the low-cost potential promised by FC,

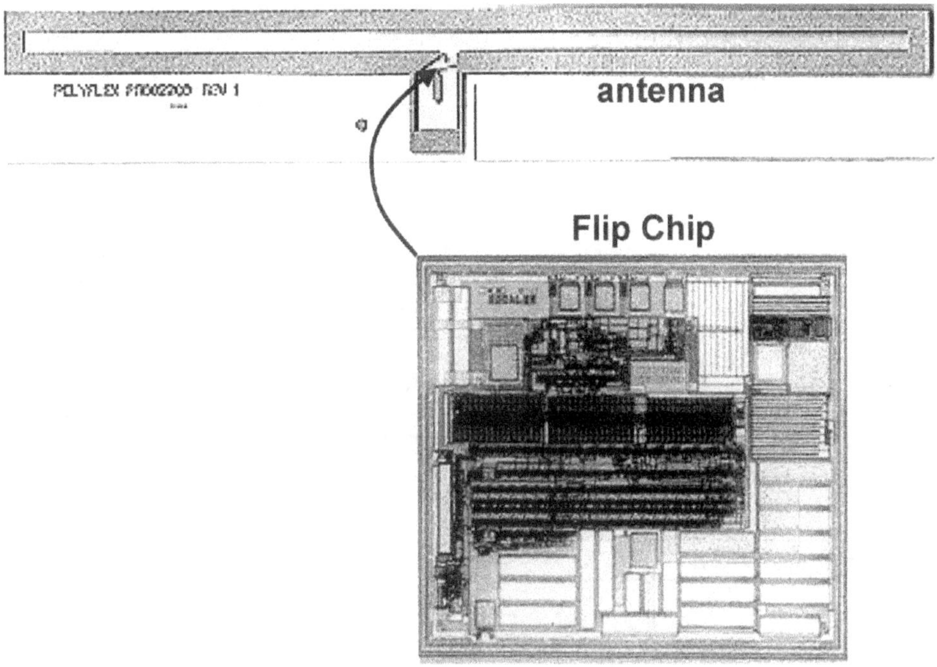

FIGURE 6.20 RFID with FC.

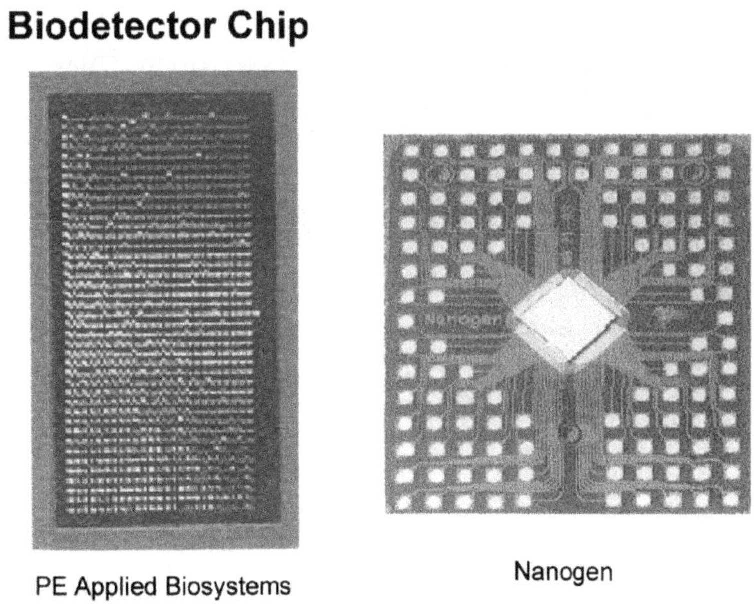

FIGURE 6.21 DNA detector, SmartFlex.

and it is certainly on the roadmap of the semiconductor industry.[50–52] And third-generation wafer-level FC will be the final enabling technology to allow FC to achieve maximum performance, manufacturability, and productivity. Just think of a chip-sized package, only 1 or 2 mils thicker than the chip, that can accommodate 10,000 I/Os and run at more than 1 GHz. Now think of the simple surface-mount assembly process, where all the necessary ingredients are right there with the chip. Solid solder, flux, and underfill are an integral part of the IC. And if this were not enough, the finished assembly would be reworkable. Should all this happen as planned, the integral flux/underfill FC would be the most important micropackage—truly the ultimate.

Although major strides are being made at a rapid pace in the FC arena, many issues impede this technology. Perhaps most important is a lack of centralized infrastructure. Major companies such as IBM, Motorola, and Amkor may have built their own infrastructure, but it is these islands of technology, often proprietary and closely guarded, that inhibit mainstream progress. Although advances have been made in bumping, much more is needed in this area, as well as in design and test. Maximum benefits from FC must start at the chip level, with area array pad layout resulting. On-chip routing of conductors to the perimeter followed by rerouting back to the center is counterproductive to say the least. The full FC infrastructure, starting with design, must be put in place to achieve the maximum payoff. This will happen, but the timetable is still unclear. What is certain is that FC will become the ultimate micropackage.

6.13 REFERENCES

1. Hayden, T., and Partridge, J., "Practical Flip Chip Integration into Standard FR-4 Surface Mount Processes: Assembly, Repair and Manufacturing Issues," in *Proceedings of the 1st International Symposium on Flip Chip Technology*, San Jose, CA, Feb. 15–18, 1994, pp. 1–7.
2. Lowe, H., "No Clean Flip Chip Attach Process," in *Proceedings of the 1st International Symposium on Flip Chip Technology,* San Jose, CA, Feb. 15–18, 1994, pp. 17–24.
3. Ono, M., T. Shiraishi, Y. Bessho, et al., "Area-Array Interconnect Using Stud-Bump-Bonding," in *Proceedings of the International Symposium on Microelectronics,* sponsored by IMAPS, San Diego, CA, Nov. 1–4, 1998, pp. 893–898.
4. Zuma, S., and Kaga, Y., "A Stud Bumping Technology Using Copper Wire," IMAPS 2nd International Advanced Workshop on Low Cost, Braselton, GA, March 13–15, 1998.
5. Oppert, T., Zakel, E., and Teutsch, T., "A Roadmap to Low Cost Flip Chip Technology and Chip Size Packaging Using Electroless Nickel Gold Bumping," IMAPS 2nd International Advanced Workshop on Low Cost, Braselton, GA, March 13–15, 1998.
6. Anhock, S., Ostmann, A., Oppermann, H., Aschenbrenner, R., and Reichl, H., "Reliability of Electroless Nickel for High Temperature Applications," in *Proceedings of the Advanced Packaging Materials,* sponsored by IMAPS, Braselton, GA, March 14–17, 1999, pp. 256–261.
7. Ostmann, A., Kloeser, J., Zakel, E., and Reichel, H., in *Proceedings of the 1996 International Flip Chip, Ball Grid Array, TAB and Advanced Packaging Symposium (ITAB 96),* Sunnyvale, CA, Feb. 14–16, 1996, pp. 152–157.
8. Simon, J., Zakel, E., and Reichl, H., "Electroless Deposition of Bumps for TAB Technology," in *Proceedings of the 40th ECTC,* Las Vegas, NV, 1990, p. 44.
9. Yoneda, Y., Kuramochi, T., Sohara, T., et al., "A Novel Flip Chip Bonding Technology Using Au Stud Bump and Lead-Free Solder," in *Proceedings of the 4th Annual Pan Pacific Microelectronics Symposium,* Kauai, HI, Feb 2–5, 1999, pp. 147–152.
10. Wolf, J., Chimiel, G., Simon, J., and Reichel, H., "Solderbumping—A Comparison of Different Technologies," in *Proceedings of the 1st International Symposium on Flip Chip Technology,* San Jose, CA, Feb. 15–18, 1994, pp. 105–111.
11. Huang, Y.-W., Collier, P., Teo, K., et al., "Wafer Bumping by Stencil Printing," in *Proceedings of the Pan Pacific Microelectronics Symposium,* Kauai, HI, Feb. 10–13, 1998, pp. 455–460.
12. Nguty, T., Riedlin, N., and Erere, N., "Solder Paste Characterization for Flip Chip Applications," in *Proceedings of the Technical Program of Nepcon West,* Anaheim, CA, Feb. 22–27, 1999, pp. 589–601.
13. Kaga, Y., and Zhang, J., "Low Cost Wafer Bumping with Maskless Process," in *Technical Proceedings of Surface Mount International,* San Jose, CA, Aug. 23–27, 1998, pp. 265–269.
14. Godin, R., "Flip Chip Bumping with High Speed Metal Jet Technology," in *Proceedings of the Technical Program of Nepcon West,* Anaheim, CA, Feb. 22–27, 1999, pp. 1489–1495.

15. Neugebauer, C. A., and J. A. Loughran, "Method of Fabricating Gold Bumps on ICs and Power Chips," U.S. Patent 4,750,666, June 14, 1988.
16. Eldring, J., Prokoph, G., Bauer, A., et al., "Flip Chip Attachment Using Mechanical Bumps," in *Proceedings of the 1st International Symposium on Flip Chip Technology,* San Jose, CA, Feb. 15–18, 1994, pp. 74–81.
17. Lyons, A., Hall, E., Wong, Y.-H., and Adams, G., "Flip Chip on Flex with Anisotropic Conductive Adhesives," in *Proceedings of Flexcon,* Sunnyvale, CA, Oct. 12–14, 1994, p. 17.
18. Keswick, K., et al., "Compliant Bumps for Adhesive Flip Chip Assembly," in *Proceedings of the 1st International Symposium on Flip Chip Technology,* San Jose, CA, Feb. 15–18, 1994, pp. 66–73.
19. Lasky, R., Morvan, Y., and Baldwin, D., "What Is Needed to Establish Flip Chip as a Standard SMT Process," in *Proceedings of Technical Program of Nepcon West,* Anaheim, CA, Feb. 22–27, 1999, pp. 1025–1030.
20. Johnson, C., and Baldwin, D., "Pre-Applied Underfills for Low Cost Flip Chip Processing," in *Proceedings of Advanced Packaging Materials,* Braselton, GA, March 14–17, 1999, pp. 73–76.
21. Ludwig, R., and Lee, N.-C., "Achieving Ultra-Fine Dot Solder Paste Dispensing," in *Proceedings of Technical Program of Nepcon West,* Anaheim, CA, Feb. 22–27, 1999, pp. 576–588.
22. Nave, J., Zakel, E., and Reichl, H., "TC-Bonding for FC-Technology on Ceramic, Silicon and Organic Substrates," in *Proceedings of the 1996 International Flip Chip, Ball Grid Array, TAB and Advanced Packaging Symposium (ITAB 96),* Sunnyvale, CA, Feb. 14–16, 1996, pp. 90–98.
23. Higashi, K., Ontani, H., Yagi, Y., et al., "Development of Conductive Stud Bump Technology for the Best Shape Design and Control," in *Proceedings of the Pan Pacific Microelectronics Symposium,* Hawaii, Feb. 10–13, 1998, pp. 449–454.
24. Baldwin, D., "Advances in Low Cost Flip Chip Process Technology," IMAPS 2nd International Advanced Workshop on Low Cost, Braselton, GA, March 13–15, 1998.
25. Ito, S., and Akizuka, S., "A New Flip Chip Packaging by Non-Conductive Resin Sheet and the Stress Management," IMAPS 2nd International Advanced Workshop on Low Cost, Braselton, GA, March 13–15, 1998.
26. Gilleo, K., "Will Flip Chip Become the Ultimate CSP?" IMAPS 2nd International Advanced Workshop on Low Cost, Braselton, GA, March 13–15, 1998.
27. Gilleo, K., "Transforming Flip Chip Into CSP With Reworkable Wafer-Level Underfill," in *Proceedings of the 4th Annual Pan Pacific Microelectronics Symposium,* Kauai, HI, Feb 2–5, 1999, pp. 159–165.
28. Gilleo, K., "New Generation Underfills Power the 2nd Flip Chip Revolution," in *Proceedings of the Pan Pacific Microelectronics Symposium,* Hawaii, Feb. 10–13, 1998, pp. 147–154.
29. Gilleo, K., and Blumel, D., "The Ultimate Flip Chip—Integrated Flux/Underfill," in *Proceedings of the Technical Program of Nepcon West,* Anaheim, CA, Feb. 22–27, 1999, pp. 152–160.
30. Crane, L., Torres-Filho, A., Yaeger, E., and Heuel, M., "Development of Reworkable Underfills, Materials, Reliability and Processing," in *Proceedings of the Technical Program of Nepcon West,* Anaheim, CA, Feb. 22–27, 1999, pp. 144–151.
31. Capote, A., "A Novel Flip-Scale Package Using Pre-Applied Multilayer Flip Chip Under-Encapsulation," in *Proceedings of the 4th Annual Pan Pacific Microelectronics Symposium,* Kauai, HI, Feb 2–5, 1999, pp. 153–158.
32. Gilleo, K., "Flip Chip 1, 2, 3: Bump, Bond and Fill," *Circuits Assembly,* June 1996, pp. 32–34.
33. Gilleo, K., "Flip or Flop?" *Circuits Assembly,* Feb. 1997, pp. 40–48.
34. Anderson, B., Bacher, B., and Gomez, M., "Development Methodology for a High-Performance, Snap-Cure Flip-Chip Underfill," in *Proceedings of the Technical Program of Nepcon West,* Anaheim, CA, Feb. 22–27, 1999, pp. 135–143.
35. Gilleo, K., and Blumel, D., "The Great Underfill Race," in *Proceedings of the International Symposium on Microelectronics,* San Diego, CA, Nov. 1–4, 1998, pp. 704–706.
36. Duck, A., "Flip Chip Underfill," IMAPS 2nd International Advanced Workshop on Low Cost, Braselton, GA, March 13–15, 1998.
37. Norris, M., "The Dispensing Process in Advanced Electronic Component Manufacturing of Ball Grid Arrays, Flip Chip and Chip Scale Packages," in *Proceedings of the Pan Pacific Microelectronics Symposium,* Hawaii, Feb. 10–13, 1998, pp. 179–185.
38. Wyllie, G., and Miquel, B., "Technical Advancements in Underfill Dispensing," in *Proceedings of the Technical Program of Nepcon West,* Anaheim, CA, Feb. 22–27, 1999, pp. 152–157.
39. Nguyen, L., Fine, P., Cobb, C., Tong, Q., Ma, B., and Savoco, A., "Reworkable Flip Chip Underfill—Materials and Processes," in *Proceedings of the International Symposium on Microelectronics,* IMAPS, San Diego, CA, Nov. 1–4, 1998, pp. 707–719.
40. Cowburn, A., et al., "Flexible Circuit Solutions for Advanced MR Head Disk Drives," in *Proceedings of Flexcon,* Sunnyvale, CA, Oct. 12–14, 1994, pp. 1–6.

41. Cowburn A., et al., "Flex Circuit Requirements for Flip Chip Attach," in *Proceedings of the 1996 International Flip Chip, Ball Grid Array, TAB and Advanced Packaging Symposium (ITAB 96),* Sunnyvale, CA, Feb. 14–16, 1996, pp. 75–79.
42. Masumoto, T., et al., "Stud Bumped Flip Chip on Flex (FOF) in Hard Drive Applications," in *Proceedings of the 1996 International Flip Chip, Ball Grid Array, TAB and Advanced Packaging Symposium (ITAB 96),* Sunnyvale, CA, Feb. 14–16, 1996, pp. 59–63.
43. Jang, S.-Y., et al., "Study on the Under Bump Metallurgy of Electroplated Eutectic Pb/Sn Solder Bumps on Organic Substrate," in *Proceedings of the 4th Annual Pan Pacific Microelectronics Symposium,* Kauai, HI, Feb 2–5, 1999, pp. 173–179.
44. Koopman, N., et al., "Solder Flip Chip Developments at MCNC," in *Proceedings of the 1996 International Flip Chip, Ball Grid Array, TAB and Advanced Packaging Symposium (ITAB 96),* Sunnyvale, CA, Feb. 14–16, 1996, pp. 64–74.
45. Rosson, J., Clawson, R., and Ihms, D., "Flip Chip Underfill Characterization methods: Developing a Test Methodology for Success in the Harsh Automotive Environment," in *Technical Proceedings of Surface Mount International,* San Jose, CA, August 23–27, 1998, pp. 295–302.
46. Price, D., and Corbett, S., "High Volume Flip Chip on Flex with Conductive Adhesives," in *Proceedings of the Technical Program of Nepcon West,* Anaheim, CA, Feb. 22–27, 1999, pp. 1502–1508.
47. Gilleo, K., Boyes, B., Corbett, S., Larson, G., and Price, D., "High Volume, Low Cost Flip Chip Assembly on Polyester Flex," *Circuit World,* 25 (2), Feb. 1999, pp. 11–17.
48. Roybal, T., "Developing Aerospace/Military Flip Chip Technologies," IMAPS 2nd International Advanced Workshop on Low Cost, Braselton, GA, March 13–15, 1998.
49. LeClair, T., Harper, A., Graham, S., and Ackley, D., "Flip Chip Interconnect of DNA Chip Devices," in *Proceedings of the International Symposium on Microelectronics,* San Diego, CA, Nov. 1–4, 1998, pp. 732–736.
50. Surface Mount Council Report, "Status of the Technology Industry Activities and Action Plan," San Jose, CA, Aug. 23–27, 1998.
51. "The National Technology Roadmap for Semiconductor Technology Needs," Semiconductor Industry Association (SIA), San Jose, CA, 1997.
52. Werner, R., Frear, D., DeRosa, D., and Sorongon, E., "Flip Chip Packaging," in *Proceedings of the Advanced Packaging Materials,* IMAPS, Braselton, GA, March 14–17, 1999, pp. 246–251.

CHAPTER 7
OPTIONS IN HIGH-DENSITY PART CLEANING

Steve Randolph and John Sanders
Speedline ACCEL

7.0 INTRODUCTION

While high-density packages such as flip chip (FC) and ball grid array (BGA) devices deliver substantial input-output (I/O) resources, their close dimensions introduce substantial cleaning challenges. These devices may have interconnect pitch dimensions of 250 μm and standoff distances in the range of 50 to 100 μm. These packages must be absolutely free of contamination in order to meet performance and dependability standards. However, contaminated interconnect areas are only accessible from the periphery, making it very difficult to reach and clean flux residues from the very center of a package.

As with other soldered packages, high-temperature reflow of high-density devices can create charred and caramelized flux residues that are difficult to remove. In addition, any residual solvent, water, or other manufacturing by-products left after BGA or FC cleaning will lead to ionic contamination and corrosion.

Process residues also interfere with subsequent underfilling, resulting in voids that can promote moisture collection, overheating, and part failure. A cleaning system must remove all flux residue, solder balls, ionic contamination, oils, and foreign matter, as well as chemical cleaning agent residues.

Manufacturers have three primary alternatives for postreflow cleaning of flux residues from large packages with fine-pitch interconnects and minimal clearances. These are in-line spray, ultrasonic bath, and centrifugal cleaning. Table 7.1 summarizes the primary characteristics of each of these cleaning methods.

7.1 IN-LINE CLEANING

In-line spray cleaners are extended, conveyorized systems that direct cleaning solvents against parts to be cleaned using vertically oriented spray nozzles. Devices being cleaned pass through wash, prerinse, rinse, and drying zones on a continuous conveyor. For difficult cleaning applications, nozzle count and spray pressure may be increased, and extra in-line cleaning stages may be added to achieve the desired results.

As additional stages are added to increase cleaning dwell time, the system footprint is extended. In-line spray systems are good for removing flux residues and oils on exposed surfaces in a continuous production setting but generally are not effective in penetrating the close tolerance and hidden gaps of FC assemblies.

TABLE 7.1 FC and BGA Postreflow Cleaning Options

Cleaning method	Advantages	Disadvantages
In-line spray	A continuous, integrated manufacturing process	Minimal cleaning penetration around die, large equipment footprint
Ultrasonic	Relatively effective cleaning	May damage delicate parts, long-term reliability concerns, typically off-line batch
Centrifugal	Very effective cleaning process in a small floor space footprint, short cycle times, complete cleaning, total drying	Batch, off-line process

7.2 ULTRASONIC CLEANING

Batch-style ultrasonic cleaners place parts to be cleaned in a solvent immersion bath, where ultrasonic energy aids in flux removal by cavitation of the liquid, which varies according to fluid density and the frequency and amplitude of the excitation (Fig. 7.1). This process is superior to in-line cleaning for cleaning complex parts. However, in the case of FC components, ultrasonic energy may not be able to fully penetrate the gap and reach hidden contaminants. In addition, this process is relatively ineffective at removing solubilized contaminants from the confines of high-density packages. Mechanical resonance induced during ultrasonic cleaning can cause microfracture of delicate parts, thereby degrading long-term reliability.

7.3 CENTRIFUGAL CLEANING

Centrifugal cleaning, another batch process, offers an important performance advantage over other FC and BGA cleaning methods: the ability to direct the solvent laterally around and between die terminations under pressure. A centrifugal cleaning system holds devices in fixtures so that the component edges are impinged directly by solvent. A robot arm assembly containing fixtured parts is lowered into a cleaning solvent bath in a sealed process chamber and rotated alternately in clockwise and counterclockwise directions. This causes the solvent to be forced through the package in a direction parallel to the plane of the substrate, driving it under and between shadowed components. Contaminants are solubilized, suspended in solution, and driven off (Figs. 7.2 and 7.3).

7.3.1 The Centrifugal Cleaning Sequence

This cleaning process combines immersion in a semiaqueous or traditional solvent with agitation and solvent flow resulting from the centrifugal force of rotation. Because the chamber is sealed and nitrogen inerted, the solvent temperature can be raised safely to a level near the flash point, and this reduces surface tension and lowers viscosity values for increased solvent action.

As immersed circuit assemblies are rotated, the spaces beneath package features and the force vectors acting on them are in the same plane. The mass of the cleaning compound together with Coriolis forces produced by rotation propel the solution into the tightest spaces to extricate entrapped flux.

Periodic reversal of rotation ensures that liquid flow is established in every direction so that flux is rapidly solubilized and flushed away. Since both the fluid and packages are in constant motion, solvent action is continuous. In contrast, solvent action in a spray cleaning system is essentially limited to the portion of a device being impinged by the solvent spray (Figs. 7.4 and 7.5).

After the centrifugal solvent washing step, cleaning solution is drained from the sealed chamber, and a distilled water or virgin solvent rinse spray is introduced as the spin rate of the fixture suspended

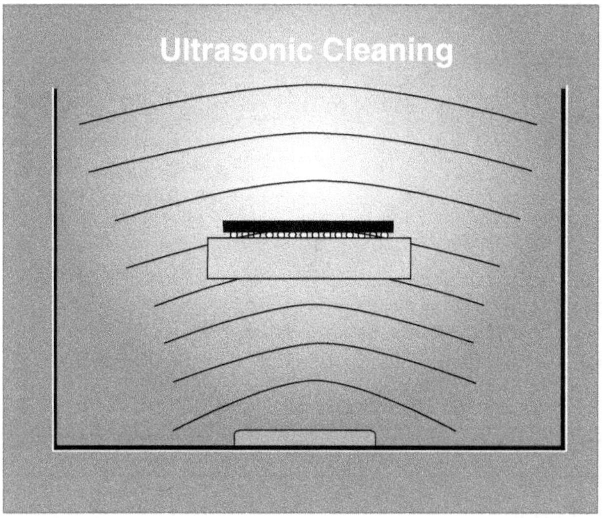

FIGURE 7.1 Ultrasonic cleaning. The energy created by an ultrasonic cleaning system displaces surface contaminants effectively but may not penetrate the hidden recesses that characterize devices that have high interconnect density.

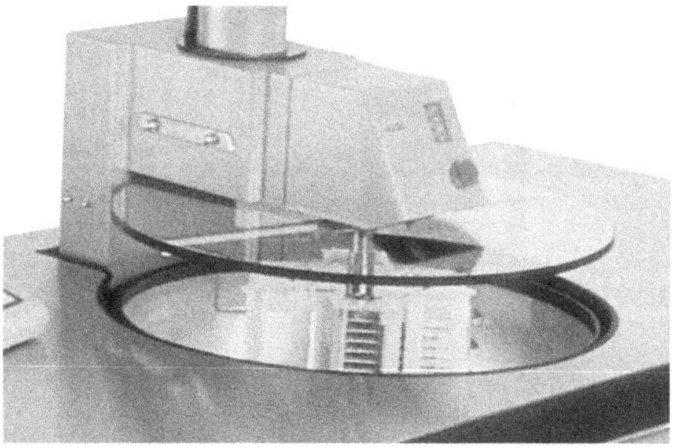

FIGURE 7.2 The Speedline ACCEL MicroCel centrifugal cleaning system. The unit uses fixtures for BGA and FC devices that are mounted on a robot arm and rotated in the cleaning chamber.

from the robot arm is increased. The rinse spray orientation is lateral so that rinse water also impinges on the edge of packages and flows between interconnects. Rotation of assemblies on the robot arm further enhances penetration of the rinse agent.

The spray rinse step is carried out in an empty chamber so that cleaned parts are not recontaminated with residue from the rinse water. When rinsing is complete, parts are dried by increased centrifugal rotation, accompanied by high-volume flow of heated air or nitrogen.

Products cleaned in a centrifugal cleaner generally are held in place and secured to the rotating robot arm head using universal adjustable fixtures, standard fixtures, or custom fixtures that are specific to a product. Manufacturers generally provide hardware to handle devices, including small circuit modules, wafers, singulated packages, Auer boats, magazines, cassettes, and JEDEC trays.

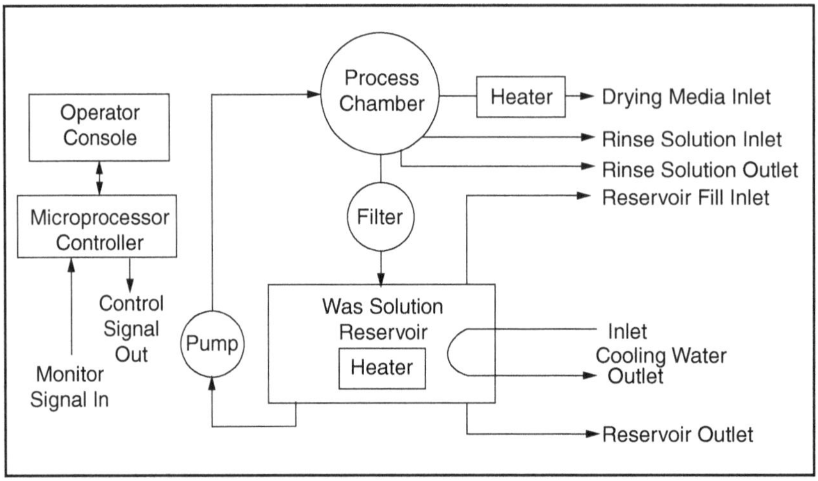

FIGURE 7.3 Machine schematic.

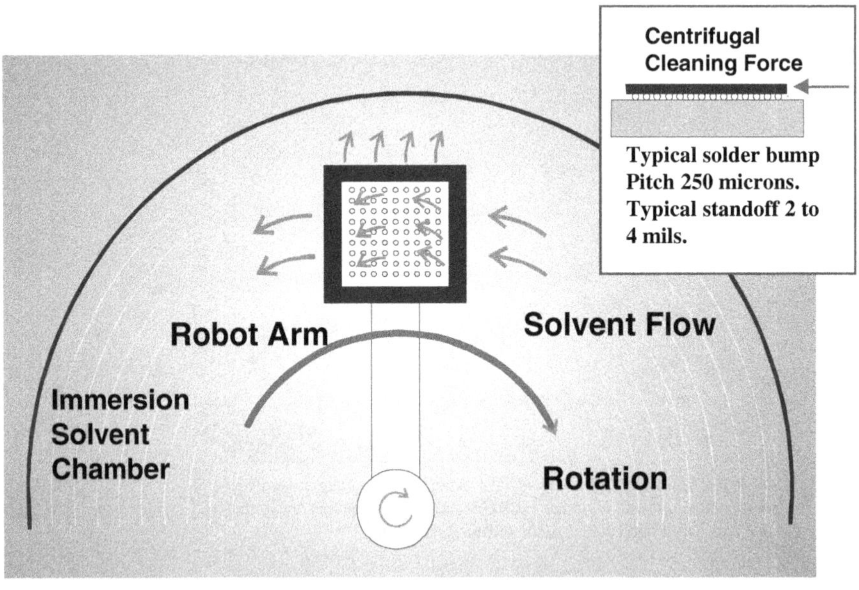

FIGURE 7.4 Movement of fixtured assemblies through the solvent bath creates positive flow of solvent between interconnections. Full coverage is ensured by alternating the direction of rotation during a cleaning cycle. Multiple components are mounted on the centrifugal system robot arm.

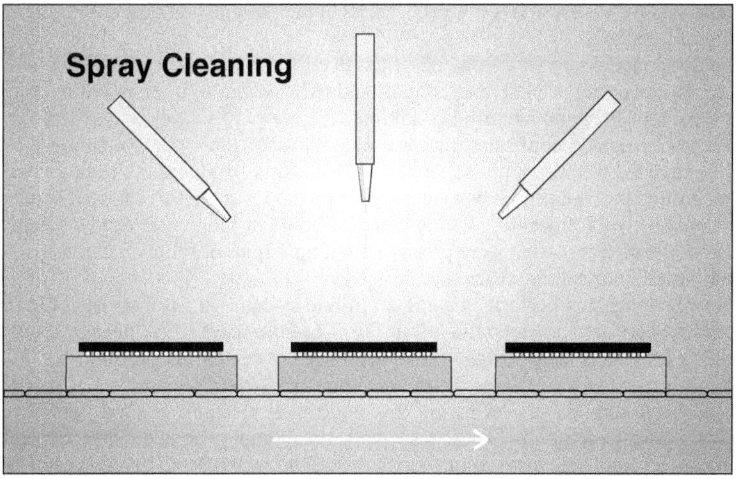

FIGURE 7.5 A spray cleaning system. This is effective for conventional circuit assemblies but may not be capable of removing flux residues and contaminants from FC or BGA assemblies.

Some centrifugal cleaning systems incorporate automatic, closed-loop, and integrated wastewater treatment. When using a solvent that will phase separate from water, such systems automatically extract it from the rinse water and return the solvent to the wash reservoir for reuse. Used rinse water may be processed through a multistage purification process that includes microbial control, microfiltration, carbon adsorption, and mixed-bed deionization to restore the water to its original level of purity. No drain or external water treatment is required for such systems, and the entire cleaning and fluid conditioning process takes place in the compact cleaning system housing.

A centrifugal cleaning system may require process recipe adjustment for optimal performance depending on part complexity, the solvent in use, and the nature of contaminants to be removed. Adjustable process variables include centrifugal rotation speed; the dwell times of wash, rinse, and dry cycles; and solvent and rinse agent temperature.

While centrifugal cleaning is a batch process, this cleaning method generally can keep pace with a conventional manufacturing line. For example, the ACCEL MicroCel centrifugal cleaning system can process hundreds of devices per cleaning/rinsing/drying cycle, which consumes approximately 7 to 15 minutes. This centrifugal system is compatible with semiaqueous, aqueous, alcohol, and traditional cleaning solvents.

A longer cycle time in a centrifugal cleaning system requires no additional floor space, unlike conveyorized systems, where space limitations may impose constraints on system size and thus the level of cleanliness that can be achieved. A closed centrifugal cleaning process provides good odor containment because cleaning, rinsing, and drying all take place in a sealed process chamber. With centrifugal cleaning, there is no aerosol phase, as with a spray system, no solvent drag-out, and no objectionable odors because the solvent is never exposed to the atmosphere.

The primary solvent factors to be considered for cleaning close-tolerance assemblies are solvency for the flux/contaminants in question, solvent surface tension, vapor pressure, flash point, safety/toxicity, and cost. Because of the sealed nature of centrifugal cleaning, concerns related to solvent flash point, safety/toxicity, odor, and waste disposal are minimized.

7.3.2 Solvent Performance

An evaluation of centrifugal cleaning by Texas Instruments illustrated the cleaning performance that could be achieved with centrifugal energy. This component manufacturer tested flux-removal efficiency using square glass coupons sandwiched to glass plates with clearances of 2, 5, and 10 mils.

The test contaminant was alpha 321 flux, dyed for visual inspection. This material was injected into the spaces between plate and coupon and then heated to a temperature that simulated reflow levels typical of FC manufacturing conditions.

This test sequence confirmed that average levels of ionic contamination on cleaned FC parts could be reduced to 0.0 μg per square and visible flux residue totally eliminated on inspection at 30× magnification. A hydrocarbon cleaning compound was the solvent used in the centrifugal cleaning system test, with processing taking place 24 hours or more after reflow. Results were compared with spray-cleaned parts that were processed within 1 hour of reflow and thus had reduced potential for consolidation of contaminants.

Spray cleaning was performed using an in-line system with Dupont HCFC 120 solvent. Results from this process are shown on the left in Fig. 7.6. Flux removal is apparent around the edges of all coupons, with results proportional to the clearance between glass surfaces.

The in-line-spray test plate (left side of figure) illustrates the amount of flux remaining beneath closely spaced components following this traditional cleaning method, which relies on capillary force for solvent penetration. (These tests made no attempt to simulate flow impediments such as the fine-pitch leads and area array connections that could be expected on devices of this size.)

Comparable glass plates were prepared for testing in an ACCEL MicroCel centrifugal cleaning system using alpha EC-7R semiaqueous solvent. The samples on the right in Fig. 7.6 show the amount of flux remaining following this cleaning method.

These examples graphically illustrate the improved defluxing efficiency that can be accomplished using solvent penetration propelled by centrifugal force. Additionally, the centrifugal extraction of entrapped rinse water promotes rapid drying and contributes significantly to overall throughput in aqueous and semiaqueous processes.

7.3.3 Flux/Solvent Performance

There has been an evolution in cleaning solvents for electronics applications over the past several decades. Hydrocarbon solvents were the first materials used for cleaning, and while such fluids were effective, concerns about flammability and workplace safety eventually led to adoption of nonflammable chlorinated solvents. These materials, also effective solvents, were identified eventually as potential carcinogens and replaced with chlorofluorocarbons (CFCs), which were welcomed because of their solvency, nonflammability, low vapor pressure, and favorable toxicity properties.

CFCs in various forms began to predominate in many industrial applications, and by the late 1980s, consumption of materials such as CFC-113 and 1,1,1,-trichloroethane reached nearly 1 billion pounds per year in the United States alone. However, scientists eventually discovered that the

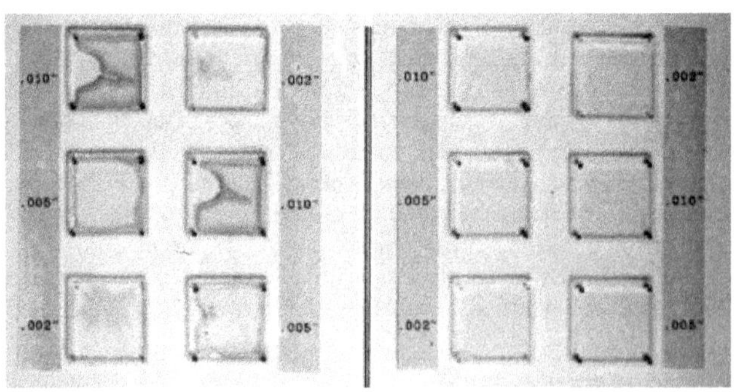

FIGURE 7.6 Glass test plates cleaned with in-line (*left*) versus centrifugal (*right*) cleaning methods.

favorable properties of CFCs came at the expense of stratospheric ozone depletion. In the mid-1990s, worldwide environmental restrictions were placed on these ozone-depleting compounds.

Today, the challenge in developing non-ozone-depleting solvents is to achieve a balance in solvent effectiveness, environmental properties, nonflammability, and safety/toxicity properties. There are a dozen or more flux formulations that can be considered for FC and BGA assemblies today, and each presents its own cleaning challenges. In the United States, the primary solvent categories used in electronics are terpenes (various citrus-derived semiaqueous solvents), long-chain alcohols, and petroleum-hydrocarbon derivatives.

Another change affecting electronics manufacturing today is the ban on use of lead-based solders in electronic manufacturing. The new lead-free solders require higher reflow temperatures, and this has created a need for modified paste/flux formulations, which, in turn, call for special solvent properties.

Manufacturers may find it useful to test the performance of the various solvent options with specific assemblies. Factors to be considered are cleaning performance with selected solder/flux materials, solvent consumption, foaming potential, environmental impact, odor, toxicity, and water treatment/disposal issues.

It may be appropriate for certain low-cost or short-term electronics applications to eliminate cleaning of high-density devices. However, the cost of effective cleaning is an incidental amount in terms of the overall manufacturing investment, and the removal of flux residues and manufacturing contaminants yields substantial benefits in package quality, reliability, and in many applications, electrical parametric performance.

7.4 SUMMARY

Centrifugal cleaning produces results that meet or exceed those of other cleaning technologies for high-density devices such as FC and BGA packages. For example, typical cleaning results measured with the ACCEL centrifugal cleaning system include ionic contamination of 0.0 $\mu g/in^2$, surface insulation resistance greater than 10^{14} Ω/in^2, and no visible residue at 30× magnification.

Centrifugal cleaning technology lends itself to a range of applications and is compatible with virtually all cleaning agents. Thus a system obtained for one application and solvent can be converted readily to another application, prolonging the useful life and productivity of the equipment. Since all processing is confined to a closed chamber, liquid storage and consumption are minimized, along with evaporative and drag-out losses.

FCs, BGAs, and other large, fine-pitch packages present some of the most difficult cleaning challenges in electronics manufacturing. However, with an appropriate cleaning solvent and a properly designed and programmed centrifugal cleaning system, it is possible to deliver contamination-free devices ready for subsequent assembly steps. The centrifugal cleaning method has been shown to support product quality, providing ease of operation, low ownership cost, and environmental compliance.

CHAPTER 8

MEMS PACKAGING AND ASSEMBLY CHALLENGES

Ken Gilleo
Cookson Electronics

8.0 INTRODUCTION

Microelectromechanical system (MEMS) may just be the ultimate enabling technology for the micro-level integration long sought for advanced products such as system-on-chip (SOC). MEMS devices operate using many forms of energy and sources of data. These include mechanical kinetic forces, light waves, sound waves, radio waves, and conventional electrical energy. This technology has been applied for complex molecular detection, a variety of analyzers, motion sensors, fluid pumping, radio signal manipulation, a host of medical applications, and photonics. When photonics or optics is added, the term *micro-optical-electromechanical systems* (MOEMS) can be used, although some prefer to use *optical MEMS.* The most extraordinary feature of MEMS is that all the diverse functions can be crafted into a single chip using modified semiconductor fabrication techniques. MEMS therefore fits the giant semiconductor infrastructure, but even older, lower-density fabrication facilities can be used. While integrated-circuit (IC) logic devices have given us the *brain,* MEMS adds the *eyes, nose, ears,* and *new senses* beyond the limits of human range. MEMS has output in the form of information that can be electronic or photonic. However, more important, MEMS also produces control functions. This adds the *muscle, hands,* and *fingers,* allowing MEMS devices to physically move internal elements and nearby objects. The *merging of motion, sensing, computation, and control* appears to represent the highest level of technology yet achieved.[1]

While packaging the versatile MEMS devices is a major challenge, old and new solutions are being deployed. Most devices are housed in costly hermetic packages, especially optical MEMS systems. Perhaps we will be able to move to a more traditional type of packaging in the future. Some less complex optical products have already been packaged in a ball grid array (BGA) format. For example, the optoBGA (iC-Haus GmbH, Bodenheim, Germany) is described as a family of optical sensors made as a micro-BGA having a glass cover.[2] The glass is placed onto the chip, which is wire bonded to a BGA substrate. The assembly is then encapsulated so that no material covers the glass, as shown in Fig. 8.1.

Many MEMS chips are hermetically sealed to exclude oxygen and moisture that can cause wear and friction problems. Most optical devices are also hermetically sealed because oxygen can degrade lasers and water can fog optics. They use atmosphere control agents such as getters (molecule-specific scavengers). Wafer-level (0-level) hermetic packaging, now under development, also may be an answer, bringing reduced cost, enabling high-volume manufacturing, and delivering the high level of protection needed. Some popular MEMS and MOEMS packaging strategies will be discussed.

However, we also face the challenge of MEMS assembly. While many packages are hand assembled, will we standardize on surface-mount technology (SMT)? Moving to SMT assembly appears

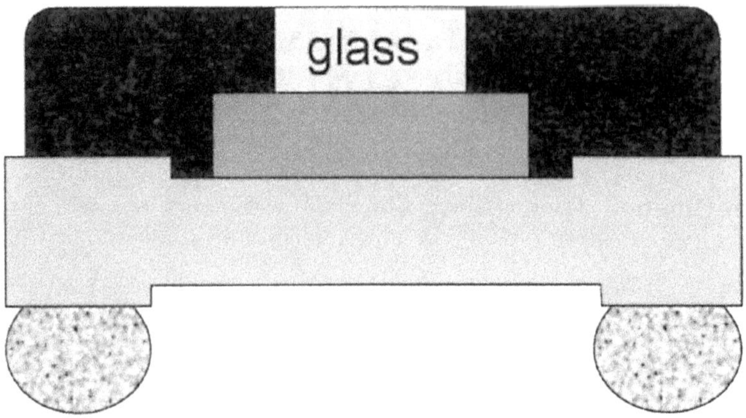

FIGURE 8.1 OptoBGA. *(iC-Haus GmbH.)*

to be a given for the high-volume markets, but there will be new rules for MEMS primarily because of the need to sense the external environment. MEMS motion-sensing products, such as automotive accelerometers for air bags, must be oriented correctly because they are highly directional. How important is package skew and tilt, and can the solder fillet even change MEMS sensitivity? We will first examine MEMS packaging and then focus on assembly issues. There may be more questions than answers at this early and evolving stage of MEMS.

MEMS may appear to be a paradox because products, mostly sensors, have been around for two decades, while forms of this technology are still emerging and in developmental form. Government and university laboratories are still working on basic problems. Sandia National Laboratories, the University of Colorado, and other research organizations have built truly remarkable devices. One example of a fascinating device is a chip-sized spectrophotometer that can detect and measure atmospheric gases from miles away. Microscopic pumps and valves also have been built that may be used for medical applications including drug delivery. Devices are so small that they can be injected into the body. The day may come when microrobots, or "nanobots," travel through the body to clear arteries and make repairs, borrowing a scene from the classic 1960s science fiction movie *Fantastic Journey.* Yet MEMS has been quietly commercialized as motion sensors for airbags and ink jet cartridges for printers. However, optical MEMS is now powering digital projectors using sophisticated micromirror arrays. These optical control modules are now being tested in cinema projectors where over 1 million miniature mirrors "paint" the movie in vivid color. Figure 8.2 shows scanning electron micrographs of the extraordinary range of MEMS devices from Sandia National Laboratory.

8.1 BACKGROUND

Electronic chips use one primary source of energy—electricity. ICs deal primarily with electronic signals. There is no mechanical movement because the flow of electrons is invisible and only manifests itself in electrical output or the emission of light in some of the electro-optical (EO) devices such as light-emitting diodes (LEDs). MEMS can represent a very high level of integration because many dissimilar kinds of functions can be combined and operated using all kinds of signals, not just electrical signals. This unification of functions includes motion, light, sound, electromagnetic radiation, and high-level analysis of external input. Computation, analysis, and central control of these I/O functions results in a fully integrated system of incredible versatility. MEMS is the convergence point of many diverse technologies. Mechanics adds gears; optics introduces mirrors, filters, and guides; the

FIGURE 8.2 MEMS from Sandia.

electrical field delivers motors; chemistry adds molecular detection; and general electronics adds radio computation and power to this new microworld. And we can expect biotechnology to enter the MEMS arena soon. The key to MEMS is that the technology blends and integrates complete functional systems. It synergistically merges subsystems into a fully integrated, self-contained microcosm that once only existed in the macro world. How remarkable that these once-isolated technologies can converge into the microscopic world of silicon using wafer-level mass processes.

However, there are many major challenges in nearly every area. Specialists in this field insist that MEMS fabrication, packaging, and assembly are the greatest challenges our industry has ever faced. Not only are the advanced MEMS devices small and highly complex, many must communicate with the outside world by direct-path nonelectronic signals. The burden placed on packaging and assembly is to provide electromechanical connections that do not interfere with or alter sensing and control function. Later, fiberoptic connections will be added. This challenge is new because assembly has dealt mostly with electrical signals. The additional I/O of motion and photonic and molecular "signals" adds a new twist.

Some MEMS devices send and receive light beams; others detect specific molecules, including biological analysis such as DNA. The merging of motion, sensing, and computation most certainly represents a new level in technology that is still embryonic. Surprisingly, this new technology has been around for nearly three decades but at a simple level such as motion sensing. More recently, however, optical MEMS, also called MOEMS, has become a popular area. The Internet is the new and powerful driving force. There are solid and durable opportunities in communications hardware. The modern Internet is really a fiberglass highway, a system of photonics light pipes. Light control is the master key to efficient photonics—the technology of communication using photons—and optical MEMS is poised to be the winner here. Just during the year 2000, the Internet giants spent billions of dollars to acquire MEMS companies. Thus, as light wave technology moves from the present long haul, or backbone, closer to the end user, we can expect the volume for MOEMS assemblies to increase and move out to the contract assemblers. The future wave may be light, not solder!

8.2 MEMS FABRICATION

Techniques have been derived mostly from IC fabrication with appropriate modifications to produce movable parts. Standard semiconductor bulk fabrication processes create the traditional electronic circuit portion. Subtractive etching produces the movable mechanical elements. Silicon dioxide can be formed as the sacrificial material that is removed in the final step to free up the movable components. The silicon or silicon nitride later becomes the mechanical or optical elements in a final "release" step. Subtractive methods are used to remove the elemental silicon, such as wet chemical etching or reactive ion milling. The sacrificial holding structure is sometimes etched away at the packaging foundry so that delicate parts are protected while the wafer is transported. Although some devices are very fragile initially, MEMS parts are more robust after packaging. However, many MEMS devices, even when packaged, need careful handling by the assembler compared with conventional electronic packages. However, the very low mass per unit volume makes the microstructures fairly robust. In addition, silicon is a very strong, high-modulus material that is considered ideal for these nanodevices. Table 8.1 lists some of the fabrication processes used for MEMS.

Table 8.2 lists some of the materials, masks, and etchants that have been used for MEMS fabrication. Note that silicon dioxide (SiO_2) can be used as both a mask and the sacrificial material. The high selectivity of the etchants makes it possible to produce the complex three-dimensional (3D) structures that are shown in Fig. 8.2.

8.2.1 MEMS Actuation

There are several ways of powering MEMS devices, but most are based on the input of electricity that also may be converted to thermal energy, hydrostatic pressure, electrostatic attraction, or magnetic energy. Interconversions are almost limitless. Lucent, for example, has developed optical devices that are powered by electricity generated from light fed into the system though an optical fiber. This can allow a photonic switch to be powered by energy sent down the fiber instead of requiring electrical lines. Since electric current is so readily available and easily transformed, electricity is the de facto standard "fuel" for MEMS. Several of the MEMS motors use electrostatic forces

TABLE 8.1 MEMS Fabrication Processes

Photolithography
Wafer bonding
Chemical etching
Plasma etching, reactive ion etching (RIE)
Ion beam etching
Ion milling
Plating
Vacuum metallizing
Electrical discharge machining (EDM)
Laser micromachining
Package assembly
Place getters
Thin film coating
Hermetic sealing
Molding

TABLE 8.2 Materials, Masks, and Etchants Used in MEMS Fabrication

Sacrificial layer	Etchant	Mask
Pure silicon	Amine base/pyrocatechol	SiO_2, Si_3N_4, metal
Si	KOH, alcohol	SiO_2, Si_3N_4
Si	Hydrofluoric, nitric, acetic acids	SiO_2
SiO_2	HF	Si, polyimide
Polysilicon	HF	Si_3N_4
Aluminum	HF	Polyimide

because they are substantial at the micro level. Very small objects have a relatively high surface area compared with mass. Static electricity forces are proportional to surface area and behave like movable capacitor plates. Very thin conductor plates and combs have very little mass. The electrostatic engines are capable of high speed and torque. These engines typically use linear or rotating combs. Conductive films and even magnetic ones also can be applied by vacuum techniques to produce MEMS engines. These systems are fairly efficient and do not generate much waste heat.

Thermal energy is very popular and is used in the high-volume MEMS ink jet pumps found in many disposable printer cartridges. The micro scale permits rapid conversion and transfer of thermal energy. Agilent has even applied the thermal concept to optical switches, where heat-activated "bubbles" alter the light beam. Dimorphic materials can be constructed that bend when heat from electricity is introduced. The bimaterial strips are popular because motion can be derived without "moving contact" surfaces that encounter friction and can have wear problems. Pneumatic and hydraulic engines also have been built that often are powered by heat derived from electricity. While all present packaging challenges, the "electric motors" are likely the easiest to deal with.

Once mechanical energy has been produced, it can be translated into useful motion. Just about every conceivable mechanical translator has been built. Many are just adaptations of the common mechanisms from our macro world such as gears and sliders. Others are somewhat unique to the microscale environment and would not be practical at a larger scale. Micromirrors that use thermal dimorphic beams are an example. Before moving into the two principal topics, packaging and assembly, let's briefly examine software specifically for this field.

8.3 SOFTWARE

One issue has been the lack of dedicated MEMS software for simulation and analysis. This gap is being filled rapidly. MEMS requires new software to bring order and product design efficiency to the industry. IntelliSense, a fast-rising MEMS star in Wilmington, MA, claims to be the only MEMS manufacturer with a full complement of software. The company's IntelliSuite CAD for MEMS includes a fabrication process database, materials libraries, 3D structure graphics, and a performance simulator/analyzer. The software is said to simulate and optimize MEMS to reduce cost, the number of prototypes, and time to market. However, new software is appearing almost monthly as the industry recognizes the importance of this field. Microcosm Technology, with well-known Cadance, has released MEMCAD. Apollo Photonics has just released an optical MEMS program simulator and analyzer called Photonics Suite. Figure 8.3 shows a micromirror simulation.

8.4 MEMS PACKAGING

Most packages are hermetically sealed to keep out particles and gases that would interfere with mechanical movement or fog and degrade optical systems.[1] The package complexity that can

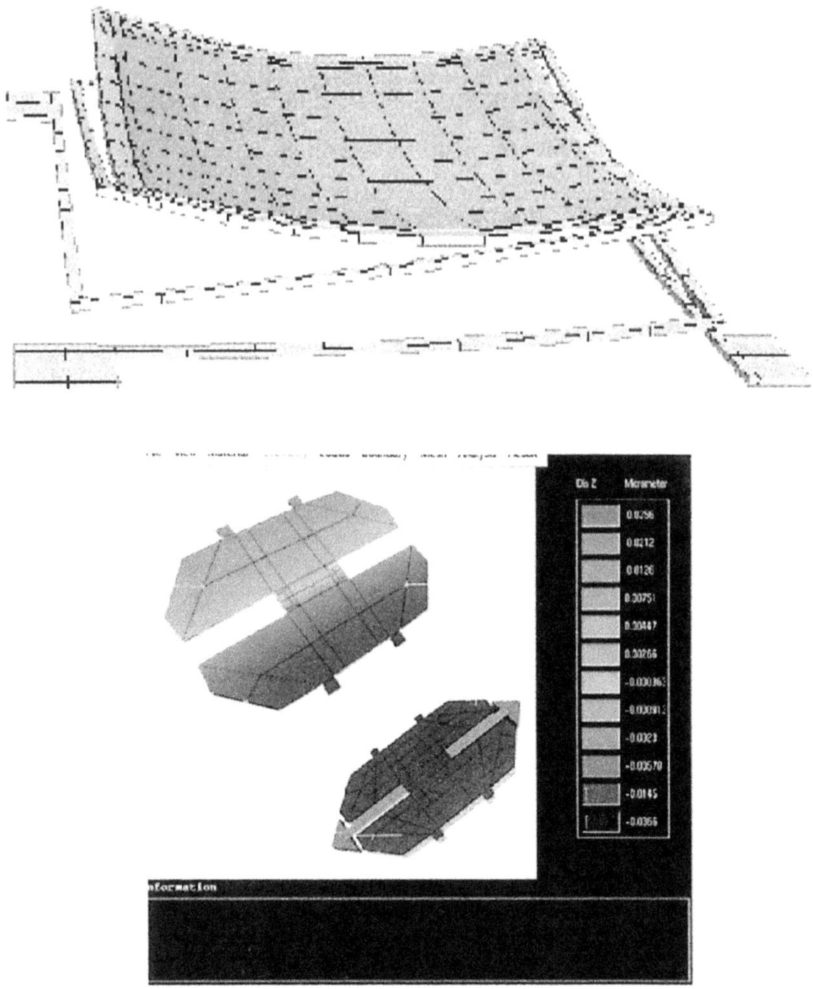

FIGURE 8.3 CAD software display of micromirror. (*From IntelliSense.*)

add assembly restrictions is highly dependent on the type of MEMS product, its functions, and the need for external access. Most MEMS devices move, but their mode and the purpose of motion determine the packaging requirements. Table 8.3 indicates some of the basic modes of motion, and several modes are shown in Fig. 8.4. Table 8.4 gives applications, I/O, and possible package types. Products with motion input and electrical output would appear to be the easiest to assemble.

The commercial MEMS industry has dealt primarily with type 1 motion devices, such as accelerometers. The industry is scaling up for two- and three-axis gyroscopes that will be used for high-volume input devices for computers, games, and all kinds of handheld products. These products are easier to handle because the package can be sealed from the outside environment. The contract assembler is likely to encounter a MEMS motion package as one of the first such packages. Packages will be both ceramic and plastic, but both will be in SMT format. Let's look at some of the old and new MEMS packaging.

TABLE 8.3 Basic Modes of Motion

1. Deformation of isolated parts: bending or twisting, no friction (accelerometers)
2. Rotating parts: friction
3. Sliding parts: friction and stiction
4. Impact: wear
5. Moving optical elements
6. Combinations of the above

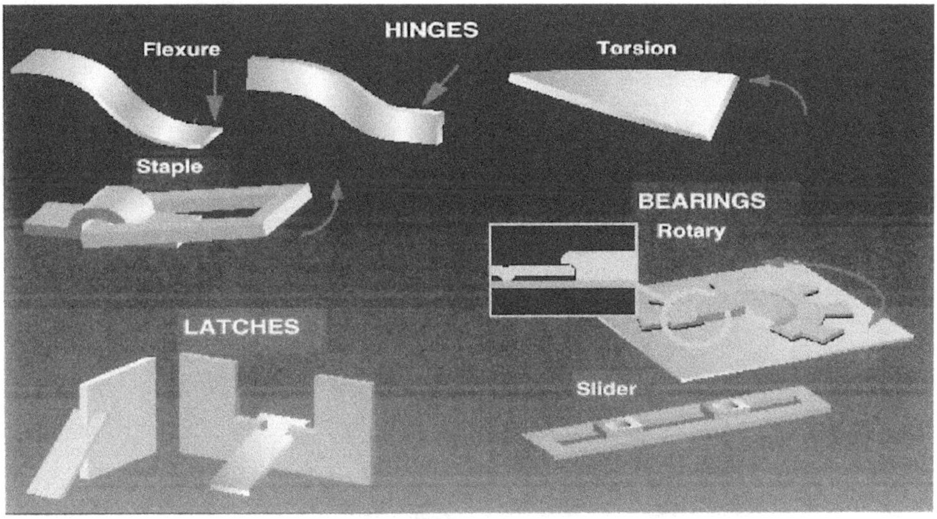

Tutorial:
MICROMACHINING IN OPTICS
Gregory A. Magel
Texas Instruments
Paper MOO1
OSA Annual Meeting
October 21, 1996

FIGURE 8.4 Modes of motion. (*From TI Tutorial.*)

8.4.1 MEMS-Specific Package Designs

Most MEMS packages started off as standard sealed hermetic containers such as a Cerdip (ceramic dual in-line) package because a ceramic or metal hermetic package offers a high level of protection but at a cost penalty. High-volume accelerometer manufacturers such as Analog Devices have converted some of these packages to SMTs called *Cerpacks.* Figure 8.5 shows the accelerometer chip. We will look at this package later to gain insight into the assembling of MEMS devices.

Now let's examine some newer packaging concepts that eventually could become standard packages that assemblers will deal with.[3] Although the MEMS package can be very application-specific, some general design concepts are emerging. The MEMS devices have a universal requirement that the "motion zone" cannot be obstructed. A protective cap over the action area can permit the component to be overmolded while still allowing unrestricted movement within the device. Once the cap is in place, standard transfer molding can be used. In some cases, liquid encapsulants may produce less stress. Liquids may be applied by automated needle dispensing equipment commonly used for some BGAs.

TABLE 8.4 Applications, I/O, and Packaging Types

MEMS device	Packaging	Input	Output
Accelerometer	EMC (epoxy mold. com'd.)	Motion	Electrical
2D and 3D motion detection	EMC	Motion	Electrical
Ink jet	Selective	Electrical	Motion
Digital mirror array	Hermetic/window	Electrical and light	Controlled light
Microspectrophotometer	Hermetic/window	Light	Electrical
DNA analyzer	Liquid access	Biological samples	Electrical
Disk drive heads	None	Magnetic	Electrical
Optical switches	Hermetic/window	Electrical and light	Electrical and light
Capacitors, tunable	EMC	Electrical/rf	rf
rf Tuners	EMC	Electrical/rf	rf
Gyroscopes	EMC	Electrical/motion	Electrical
Diagnostics (in vitro)	Open	Electrical, liquid	Electrical
Infrared imagers	Hermetic/window	IR, electrical	Electrical
Microrelay	EMC	Electrical	Electrical
Pressure sensors	Sealed with thin wall	Mechanical force	Electrical

FIGURE 8.5 Accelerometer package. (*Analog Devices.*)

Another possibility is to use flip chip (FC). Since the active surface of the device with moving parts is placed downward toward the substrate, a natural protective zone can be formed. The standoff distance, or *chip gap,* can be controlled accurately by the chip bump height. High-melting alloys or even nonfusible bumps can be used to ensure a specific minimum gap between the chip and board. The next step is to selectively dispense underfill. Normally, underfill is applied to completely fill the chip gap, but this would interfere with MEMS movement or sensing. A more viscous encapsulant, resembling damming compound, could be applied to all four edges of the chip. This sealing encapsulant, a fillet without the underfill, would then be thermally hardened. Now the package can be fully encapsulated by conventional transfer molding or by needle dispensing a liquid encapsulant. A chip access port could be added to the packaging substrate if the device must communicate with the outside atmosphere such as with a gas analyzer. A filter or semipermeable membrane possibly could be used to limit entry to only the intended molecules. Figure 8.6 shows the FC designs.[3]

A more elegant concept is to apply a microcap at the device level or, more preferably, at wafer level and then proceed to standard packaging. Several groups, including Amkor, IMEC (Belgium), and Georgia Institute of Technology, are working on or have already developed cap-on-chip, sometimes called *0-level packaging*. The cap essentially produces a silicon microhermetic package well suited for accelerometers, gyroscopes, or any other motion detectors. The present caps are made of silicon to achieve a perfect match of materials. However, the thin, flexible silicon cap may require a premolding step to prevent cave-in due to high molding pressures. Perhaps low-expansion metal caps can eliminate this problem. The alternative may be liquid encapsulation. Figure 8.7 shows the cap-on-chip, or 0-level, concept. The final package, if successfully implemented, will be similar outwardly to the common SMT plastic packages, although it is too early to know exactly what the form factors will be.

The chip scale packaging (CSP) industry also has begun to address MEMS and optical MEMS (or MOEMS). ShellCase, Inc., has developed the ShellMEMS. Windows for optical devices can be produced easily by using a transparent lid for the cap. Some package configurations use a BGA footprint. Figure 8.8 shows some of the area array designs for MEMS.

8.4.2 Packaging Atmosphere Control

The MEMS package also can consume (absorb) and emit (outgas) materials, but the atmosphere needs to be controlled within certain ranges. The atmosphere maintenance approach must be simple but effective, and a class of molecular scavengers, called *getters,* could be the best answer. Getters are agents true to their name and "get" things—primarily contaminants. Today's getters for electronic packages are very selective. These molecular scavengers are available as products that absorb gases, liquids, and solids. While the present list of commercial getters only includes moisture, particle, and hydrogen getters, additional getters can be designed as the need arises. Moisture getters use compounds that absorb and tightly bind water molecules. Zeolites and certain metal oxides are very effective. A special polymer matrix can be used as the "breathable" binder. Particle getters are made with permanently sticky polymers. This is not simply a "fly paper" material. The getter must be easy to use, maintain performance over extended time, and not produce contamination. Hydrogen getters, used to prevent poisoning of GaAs devices, are more complex. We may see the use of GaAs devices in the optical MEMS area later. Getters simply could be "printed" on the inside of the chip cap.

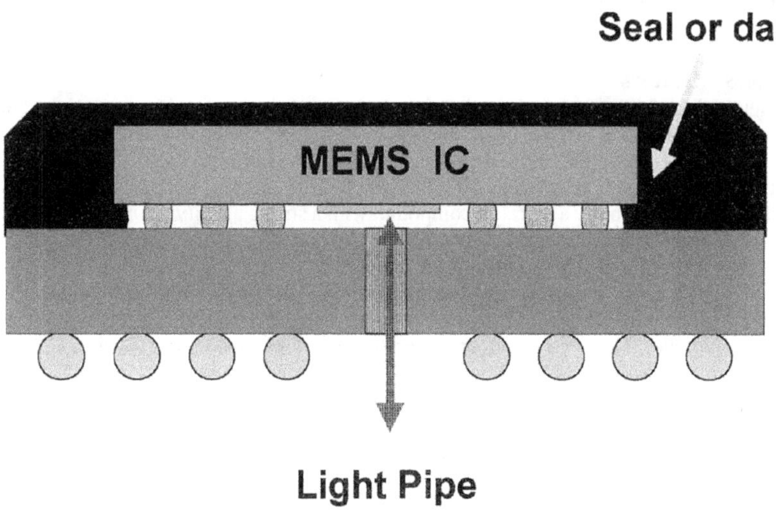

FIGURE 8.6 Flip chip MEMS packages.

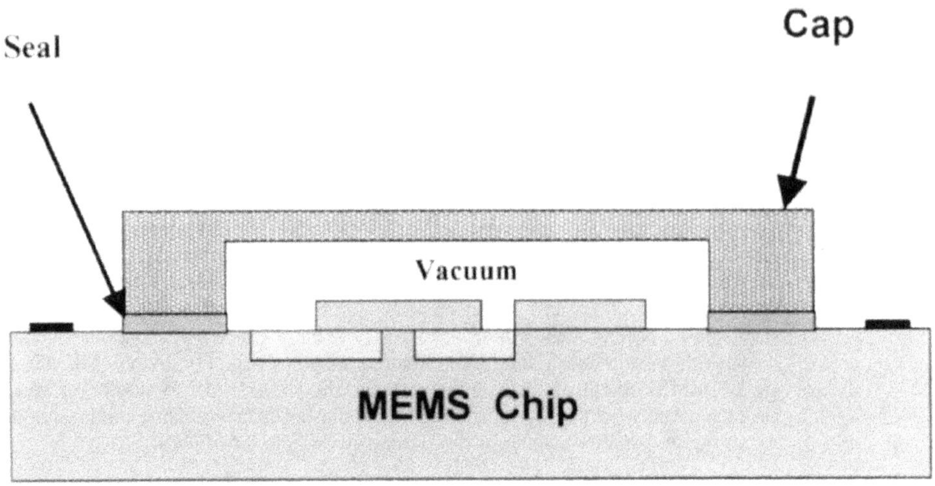

FIGURE 8.7 Cap-on-chip hermetic package.

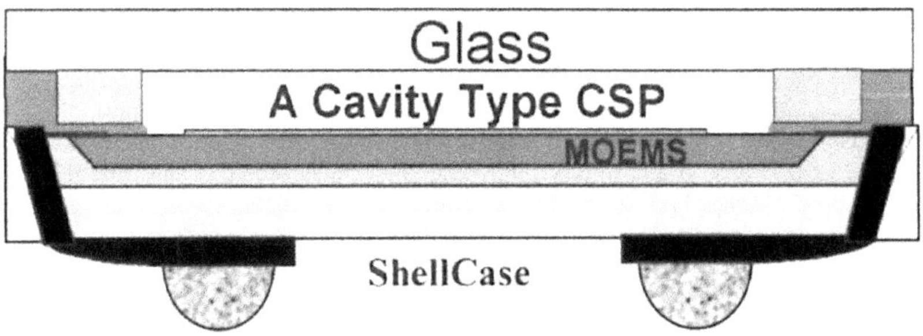

FIGURE 8.8 ShellMEMS package. (*ShellCase.*)

8.4.3 Getters

Getters have been around since the earliest phase of the electronics revolution. The early electronic vacuum tubes experienced unacceptably short lifetimes because of the oxidation of filaments and electrodes. Back in the early 1900s, it was not practical to produce a high and stable vacuum. But the life-limiting culprit was oxygen. The simple solution was to add an oxygen getter to the package. The getter was simply a metal or compound with a strong affinity for oxygen. The vacuum tube, even with a small leak, remained at a low oxygen level provided that some of the getter remained active. Now back to the twenty-first century.

Today's getter materials for modern electronic packages, while very selective, can be combined to perform multiple functions. For example, a popular product absorbs moisture and also binds microparticles. This type of getter is used for optical MEMS products because water degrades optics and particles would jamb moving parts. Moisture getters use high-capacity desiccants that absorb and tightly bind water molecules. Zeolites and certain metal oxides are very effective. A special polymer matrix can be used as a "breathable" binder, such as Staystik 415 and 482 (StayDry SD1000 and SD800, respectively). Particle getters are made with permanently sticky polymers. Creation of such polymers is a nontrivial task because outgassing must be kept low. Sticky polymer surfaces typically are produced by adding solvent or creating low-weight polymers, but this approach would produce volatiles that would be released inside the package. However, sophisticated polymer chemistry can produce high-weight, stable, but sticky surfaces. And finally, the getter must be easy to use and remain stable for years under variable conditions.

Hydrogen getters, used to prevent poisoning of gallium arsenide (GaAs) devices, are more complex. Although hydrogen-adsorbing palladium (Pd) metal can be applied to lids, a better solution is to use a polymer-bound system. This is so because the plated palladium compound has a tendency to embrittle and flake over time, leaving the possibility of unwanted particulates forming in the package. The hydrogen getter H2-3000, for example, uses palladium oxide (PdO) in combination with zeolite dispersed within a stable polymer matrix. The PdO converts H_2 gas to water that is then consumed by the zeolite, making the reaction irreversible. Incidentally, H_2 sources include the electroplating on the metal package, die-attach adhesives, and radiofrequency (rf) absorbers. Released hydrogen degrades the wire-bond interfaces of GaAs devices, reducing device performance. High-speed MOEMS optical devices could be made with gallium arsenide (GaAs) or indium phosphide (IP) that can be degraded by hydrogen. Table 8.5 summarizes the limited information available on getters.

Can getters control humidity within a specific range? This is certainly feasible, although the term *getter* may not be entirely accurate here. Specific compounds or mixtures are known that can maintain moisture within specific limits inside a sealed container. Equilibrium levels are soon reached with atmospheric water to regulate humidity. The appropriate selection of such desiccants and polymer carriers could produce atmospheric control agents for specific humidity ranges.

8.4.4 Surface Control: Friction and Stiction

As you can see, MEMS technology has its own vocabulary. *Stiction* is a form of friction in which a much higher force is required to get things moving, and the effect can be incapacitating. Once the gear, wheel, slider, or arm is unstuck, we are back to ordinary friction. Stiction goes unnoticed in the macro world, where it is a minor factor. The attractive forces in the micro world are relatively higher, and stiction is inconsequential. A high initial force may be required to start a mechanism moving if stiction cannot be overcome, and this requires bigger motors. However, microdevices have a very high surface area to mass ratio. Since friction depends on area, microdevices have stationary friction (stiction) values that can be a million times higher than friction during motion.

However, stiction can be a greater problem for even simple bending-mode devices such as accelerometers. What happens if the motion-sensing beams in the accelerometer make contact? They stick! They stay stuck! Figure 8.9 shows the stiction effect. The capacitor sensor is now a short circuit, and the device is inoperable. We cannot just pry the beams apart because the package is already sealed. But what made the beams come together? Dropping the parts or just moving them too quickly can cause the "floating" beam to move because it is designed to move with a change in motion. Thus the assembler can unknowingly "break" the part just by handling. Worse yet, the only evidence for stiction immobilization is electrical testing. While designers try to avoid MEMS devices with moving parts in contact, this is not always possible. Even motion designs with torsion beams often have stops that can stick on repeated contact.

A device that is permanently inactivated when the assembler moves it quickly is unfriendly indeed. Thus the accelerometer stiction problem is important to solve for assemblers. This will reduce handling restrictions and improve yield. There is still some debate as to what kind of atmosphere is best for reducing stiction because it can involve mechanical interlocking, atomic forces, and even chemical reactions such as hydrogen bonding. Many workers advocate the highest vacuum

TABLE 8.5 Getter Types

Getter	Form	Status
Moisture (H_2O)	Film, paste	Commercial
Particle	Film, paste	Commercial
Hydrogen (H_2)	Film, paste, thin film	Commercial
Oxygen (O_2)	Film, paste, thin-film?	Feasible
Humidity range control	Film, paste, fluid?	Feasible

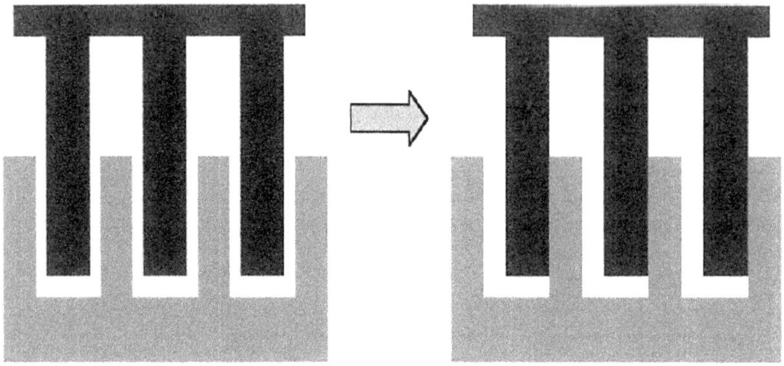

FIGURE 8.9 Stiction of an accelerometer.

possible, whereas others suggest that a small amount of moisture can reduce wear and stiction.[4] However, there are some surface chemistry complexities involved with silicon and its oxides that appear to make surface coating a logical solution. Organic coatings with low surface energy and perhaps hydrophobic properties should help.[5] The coating would need to be very thin, and this is the realm for vacuum coating technology.

8.4.5 Antifriction Coatings

Many have heard of parylene (poly-*para*-xylylene) as a special polymer film applied to substrates with a vacuum coater. The parylenes are noted for high chemical inertness, strong barrier properties, and perfect conformity to the surface. The surface tension approaches that of Teflon. A fluorinated Teflon-like version (Nova HT) has been developed recently for semiconductors that can handle over 500°C. Nova HT could be ideal for antistiction. Hopefully, work in the near future will determine if this material is a good solution to stiction and wear to deliver assembler-friendly MEMS devices.

Specialty Coating Systems, a manufacturer of chemical vapor deposition (CVD) materials and equipment, suggested that a very thin organic parylene film could solve the stiction and wear problems. Parylene, a thermoplastic polymer, has been around for years, and many assemblers are familiar with this organic CVD coating. Parylene is a high-temperature polymer film applied to substrates in a vacuum chamber by means of gas-phase polymerization that provides unusual electrical and environmental performance. This class of polymer has been used on a variety of applications, especially those involving the protection of electronic devices and circuitry. CVD is used to form an insulating thermoplastic coating with a high degree of chemical inertness, absence of pinholes, and perfect conformity to the topography of the surface. Coefficients of friction range from 0.25 to 0.33, so the lubricity is close to that of Teflon.

A fluorinated parylene, Nova HT, has been developed recently for semiconductors. Nova HT shares the unique properties of the other parylenes but offers properties that should be ideal for antistiction. The film is deposited in a molecule-by-molecule polymer process, with none of the cure-related stress that can occur with liquid polymers. There is no liquid phase, no hydraulic forces, and the coating conforms to substrate features rather than pooling or bridging in the manner of conventional liquid coatings. Free molecular dispersion of the monomer results in the development of an overlying film on all exposed surfaces, with equal thickness on inside and outside corners, on flat surfaces, and in crevices. Parylene can effectively penetrate inside surfaces through small openings.

Nova HT has a crystalline melting point above 500°C, which is at least 250°C higher than the recommended continuous exposure level for the conventional parylenes. Nova HT can be used for applications that require exposure to lasers and high-intensity lamps such as MOEMS devices with its improved ultraviolet (UV) resistance. It is particularly resistant to yellowing and physical degradation under such conditions. This advanced coating has all the useful properties of traditional parylenes, including resistance to solvents, moisture, gases, and other contaminants; high dielectric strength in very thin layers; and favorable physical and electrical properties. Its superior properties are due in part to the integration of fluorine into the parylene lattice, which results in improved polymer stability and a low dielectric constant. Hopefully, work in the near future will determine if this material is a good solution to stiction and wear.

8.5 MEMS CONSTRUCTION AND ASSEMBLY

Let's assume that the myriad of fabrication and packaging problems will be solved. The next challenge is MEMS assembly.[2] What kind special requirements and restrictions will be faced in assembly? The accelerometer and motion detectors in general are probably the best place to learn about MEMS assembly.

8.5.1 Accelerometers: Air Bag Electronics

The accelerometer is one of the earliest and most important MEMS devices. The chip typically uses a cantilevered silicon arm or a comb that bends slightly when there is a change in motion, as shown earlier. Movement is detected and translated into an electrical response. Analog Devices, a leader in this field, uses a design where a movable beam and a stationary beam form a capacitor. Motion changes the spacing and thus the capacitance to translate deceleration into an electrical signal. The system must accurately sense deceleration and send the signal that deploys the airbag. The deployment signal must only trigger under crash conditions. "Almost right" is not good enough in a life-and-death situation, as has been shown in some unfortunate incidents. The packaged product can appear very ordinary because traditional methods can be used here. This is one of the few MEMS products that can be sealed completely because motion detection does not require an opening to the outside. Still, the packaging cannot interfere with mechanical movement and must have low stress. Package stress, if present, must be predictable so that allowances can be made. The critical factor for both the packaging and assembly is *stress.* Anything that increases package stress will change sensitivity.

The accelerometer (decelerometer) must sense change in motion but not in all directions. The sensor must detect the rate of change primarily in the forward direction of the vehicle. We do not want the air bag activated if the vehicle is rear-ended or hit from the side or bounced by a pothole. Some cars are adding side air bags, but their separate sensors detect sideways motion. Anything that interferes with sensing the direction or alters the ability to detect absolute deceleration will be a problem. Let's look at how circuit assembly can affect things. Figure 8.10 shows the accelerometer diagram from Analog Devices.

8.5.2 Ink Jet

The popular ink jet chip built into most ink cartridges represents a different level of packaging challenge. The ink jet uses a MEMS chip that rapidly propels droplets when an electrical impulse is received. The chip consists of microscopic jet nozzles that discharge droplets using piezoelectric or thermomechanical pumps inside the chip. The numerous micronozzles must be kept free of packaging, but the chip-to-substrate interconnect must be protected and made robust for handling by end users. This can be accomplished by *selective packaging.* Hewlett Packard (HP) uses tape automated bonding (TAB) for the package. This flex-based package incorporates cantilevered metal beam leads that are thermocompression bonded to the IC pads. Instead of the common outer lead bond (OLB) termination, however, HP incorporates a pressure-contact connection that mates with the printer when the user installs the cartridge. This unusual and nonhermetic MEMS package therefore serves additional purposes as a 3D circuit and connector.

The IC connection zone is selectively encapsulated using robotic needle dispensing to apply liquid prepolymer sealant. Today's needle dispensing equipment allows the selective encapsulation process to be accomplished accurately, efficiently, and automatically. The deposited encapsulant is hardened thermally or with UV energy. Figure 8.11 shows the integrated flex package–connector

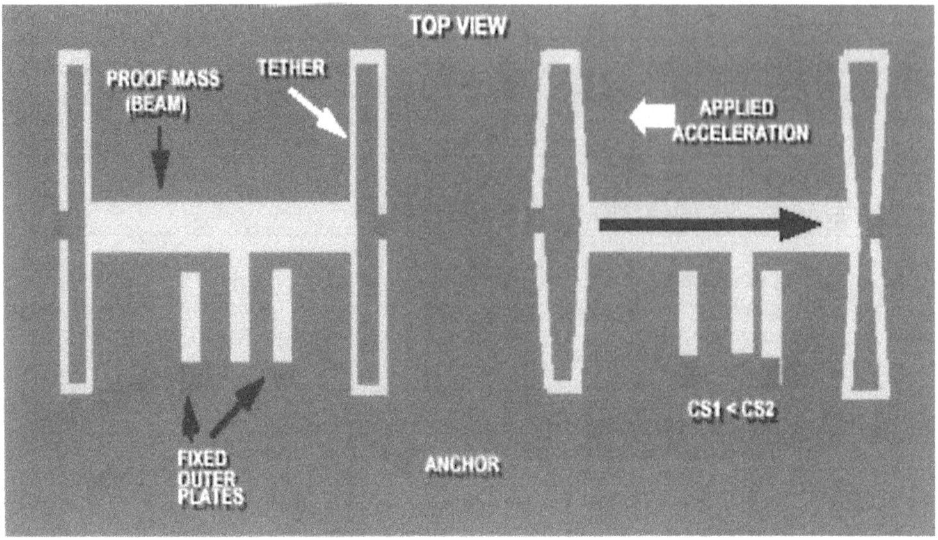

FIGURE 8.10 Accelerometer diagram. (*Analog Devices.*)

circuit, while Fig. 8.12 shows an automatic needle dispenser of the type used to selectively encapsulate a MEMS ink jet device.

8.5.3 Unique Board Assembly Issues

Analog Devices has used a Cerpack hermetic package successfully for a number of years, although considerable development was required to build a durable and reliable product. Even with a good package, assembly must be well controlled, with new rules applied. Package orientation must be correct. Skew will reduce sensitivity because the MEMS device will not point true. Hypothetically, if the package were to be skewed by 45 degrees, this would reduce the forward direction sensing by 50 percent. While it may be impossible to have this degree of skew, there is a potential for some. Component tilt is also undesirable because it also would change the effective direction of motion and reduce sensitivity. Even the solder fillet could be critical. A thicker fillet could reduce compliancy, and the sensor would receive a stronger signal because less deformation would transmit more force. Other types of MEMS products may not be as critical. On the other hand, optical systems can have more complexities for the assembler.

These are some of the more obvious assembly concerns, but there are many more that involve a change in sensing. We asked if the solder fillet was critical, and the answer is "Yes," but for mechanical not electrical reasons. Even lead compliancy plays a role in determining what the sensor "feels" in a high-impact crash. Stiffer leads will allow less deformation, and this results in a stronger signal. The assembly challenge appears to be complex.

Even the laminate characteristics must not change, but because of yet another effect. We mentioned that stress on the device, while permitted within limits, must be predictable. While Cerdip packages are popular for MEMS, there is a changeover to plastics that can increase stress and sensitivity. The manufacturer may calibrate the device during packaging as part of the test, but if the assembly process later causes unexpected changes, there will be problems. For example, board warpage and twist could be devastating. Even a change in the coefficient of thermal expansion (CTE) of the laminate would need to be taken into account. The more we delve into assembly, the more questions arise. We do not even know "what we don't know" at this point. The guiding principle comes from Amkor's Steve Anderson: "View the device, package, and assembly as an interactive system."

FIGURE 8.11 Ink jet cartridge.

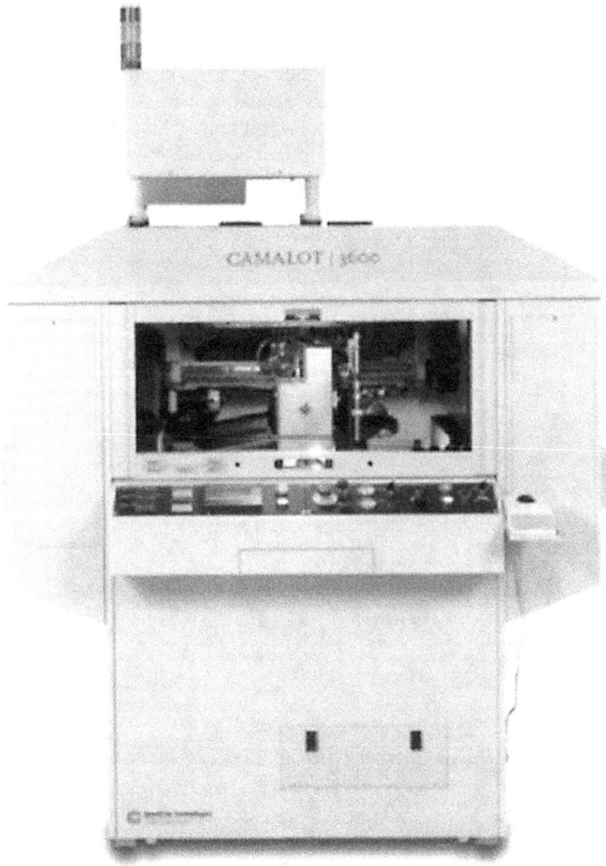

FIGURE 8.12 Needle dispensing machine. (*Speedline.*)

8.5.4 Recent MEMS Products

Camcorders with motion stabilizers use MEMS motion sensors, and Amkor is already packaging them. Again, the packaging challenge is minor for such products that can be sealed, but package-induced stress must be kept at a minimum or at least factored in. The strategy is to cap the package with silicon followed by transfer molding. Amkor points out that this area of packaging is very application-specific. Motion detectors are also being used in games.

IBM's new flash card–sized MicroDrive uses MEMS to create the world's smallest and highest-density removable hard drive (1 GB). MEMS brings the high level of precision required. The card can be used in the PC slot or in cameras, etc. Note that this is a real hard drive, not memory chips. MicroDrive is shown in Fig. 8.13.

8.6 OPTICAL MEMS: MOEMS

Optical MEMS products add one more level of complexity. The chip should be hermetically sealed, but a light path is another obvious requirement. The solution is somewhat apparent, but implementation can be a Herculean task. A light-transmissive lid or "porthole" is designed into the package. Several materials can be used. The micromirror module from Texas Instruments is one of the best examples of the packaging of complex electrophotonic products. Figure 8.14 shows the Digital Micromirror Device (DMD).[1]

The Digital Micromirror Device is very likely the most sophisticated MEMS product that has yet been commercialized and a preview of what lies ahead. The MEMS chip incorporates light beam–directing mirrors that move independently and almost instantaneously during operation. A pixel is turned "on" by pointing a mirror at a projection lens, while turning "off" involves pointing the mirror away. Let's look more closely at the optical package. Figure 8.15 shows a section of the micromirrors with some pointed "on." Large arrays are being used for digital projectors right now, but there are other applications such as digital video and optical switches.

FIGURE 8.13 IBM MicroDrive.

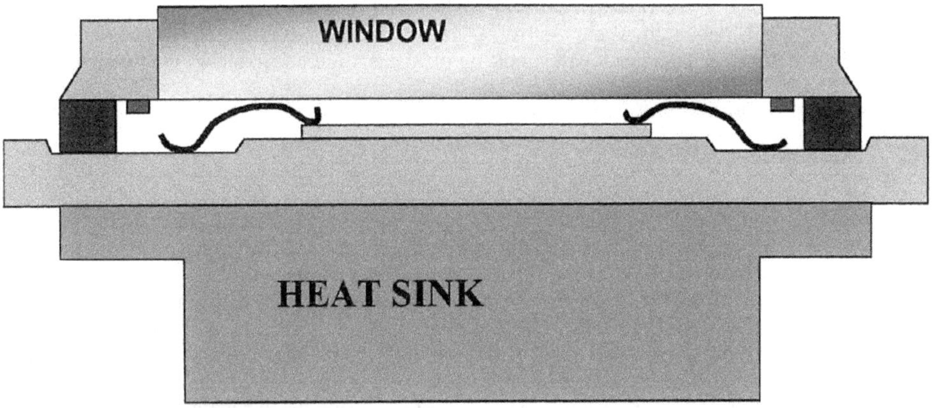

FIGURE 8.14 TI's DMD and package diagram.

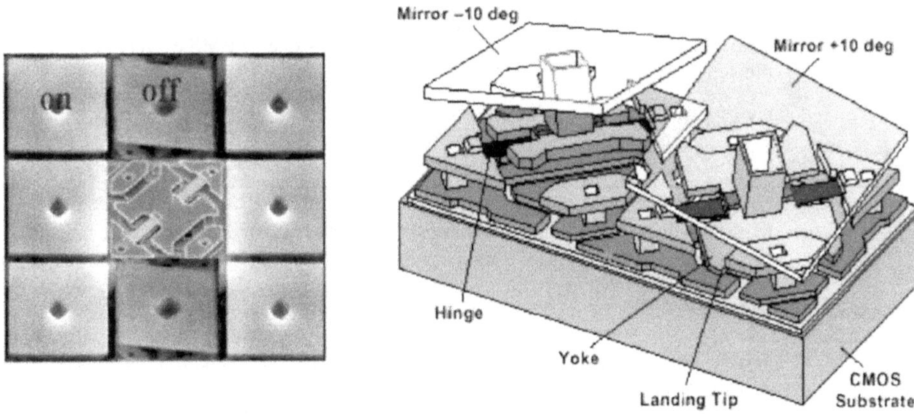

FIGURE 8.15 TI's micromirror array close up.

Most optical systems use a high-vacuum hermetically sealed package. Although many researchers recommend a high vacuum, others suggest a specific range for gases. Movable optical systems require that the package atmosphere not only start off clear but also that nothing within will later degrade lens clarity. One added problem is that materials can outgas to generate atmospheres that cause damage later, even though the initial package had a very high vacuum. Microscopic particles also can dislodge or even form during operation. How can we deal with gases and contaminants that are produced long after the package is sealed and the product is in the field? The best solution appears to be getters, which were discussed earlier.

The Internet giants, such as Cisco Systems, Nortel Networks, Lucent, Alactel, and others, have bet billions of dollars on MOEMS. Right now, the network routers and switches run optical-electronic-optical (OEO). This means that an Internet light signal must be converted to electronic format, switched, and then reconverted back to light. This adds cost and slows down the switch. The industry is intent on developing full optical switching, where photons are routed without the double conversion. One way is to use micromirrors to catch and direct the light wave. While this is easier said than done, it almost certainly will happen. While movable mirrors are the most obvious solution, there are other technologies within MOEMS that ultimately may win a piece of the light show action. So stay tuned to see who will catch the light wave. Lucent is already shipping MOEMS switches with two-axis "point anywhere" mirrors. The use of 3D mirrors greatly reduces the total number of mirrors required compared with simple on/off types. Figure 8.16 shows the two-axis optical switch mirrors.

8.7 WHY AREA ARRAY?

Will MEMS really move to area array packaging? After all, most of the packages are classic hermetic types that typically are not designed as area array or even surface mount. But it certainly is possible to design and build area array hermetic packages. Ceramic ball grid array (CBGA) and pin-grid array (PGA) are good examples of area array packages that can be hermetic and are suited for MEMS and MOEMS. Also, the micropackaging industry has already addressed the MEMS field. ShellCase has built ShellMEMS, as described earlier. Much work is also going on by packaging foundries and universities in the 0-level category, as discussed earlier.

Area array packaging, especially BGAs, bring many benefits to MEMS and could obviate several of the assembly concerns. Orientation accuracy, for example, should improve due to the highly symmetrical self-alignment. Joints would be much more controlled because BGA solder comes from the precise solder spheres. And MEMS can be expected to shrink in size while growing in complexity so that area array will be needed for the high I/O count. There is every reason to expect that area array packaging will become important for higher-level MEMS even if this does not hold for the simple sensors.

8.8 SUMMARY AND CONCLUSIONS

MEMS will be a hallmark technology for the twenty-first century. The capacity to sense, analyze, compute, and control, all within a single chip, will provide new and wonderful products during the coming decade. While package challenges are substantial, progress is accelerating. The need to control and regulate the package atmosphere will be critical. Stiction and wear problems may be solved

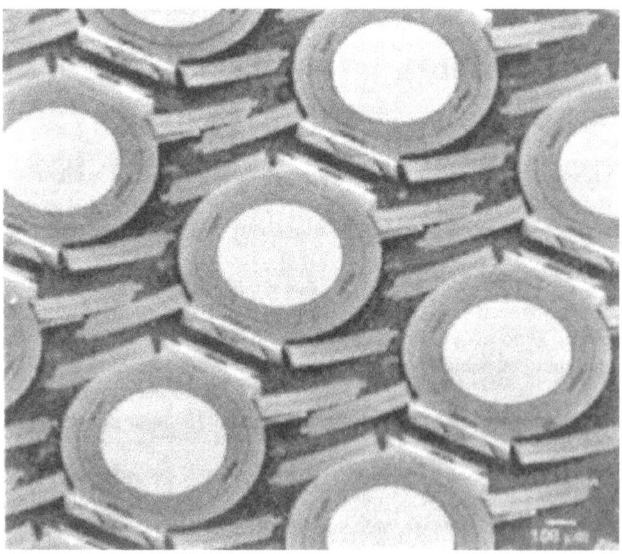

FIGURE 8.16 MOEMS switch for the Internet. (*Lucent.*)

in the future with new vacuum-applied polymers. Getters and emerging control agents appear to offer a practical and cost-effective solution today and for future generations of MEMS products. Assembly will have some different rules, but the rules will be less restrictive if package sensitivity issues can be solved. Expect to see MEMS as a high-volume component in the future, one that will usher in the new era—*beyond electronics!*

8.9 REFERENCES

1. Gilleo, K., "MEMS Packaging Solutions," *Electronic Packaging and Production,* June 2000, pp. 49, 50, 52, 53, 55, 56.
2. Quasdorf, J., et al., "BGA Goes Opto! Introducing optoBGA Packaging Technology," System Integration in Micro Electronics Exhibition and Conference, Nuremberg, Germany, June 27–29, 2000.
3. Gilleo, K., "MEMS PCB Assembly Challenge, *Circuits Assembly,* March 2000, pp. 62, 64, 66, 68, 70.
4. Miller, W. M., "MEMS Reliability and Testing," IMAPS Packaging of MEMS Microsystems Workshop, Chicago, October 23–24, 1999.
5. Martin, J., and Zhao, Y., "Micromachined Devices Packaged to Reduce Stiction," U.S. Patent No. 5,694,740, December 9, 1997.

CHAPTER 9
CERAMIC BALL AND COLUMN GRID ARRAY OVERVIEW

Marie S. Cole
IBM Microelectronics

9.0 INTRODUCTION

During the 1990s, ceramic ball grid array (CBGA) and ceramic column grid array (CCGA) packages moved from development laboratories[1] to volume factory production.[2] The steady growth in the use of these packages typically has been driven by either performance factors, process compatibility, or a combination of the two. These high-interconnection-density, surface-mount-compatible packages offer many advantages over traditional pinned or peripheral-leaded packages. CBGA and CCGA packages currently can be found in a wide range of applications spanning telecommunications and personal computers (PCs) to supercomputers. This chapter gives an overview of the package structures, range of offerings, established infrastructure, performance attributes, and package interconnection processes.

Increasing complexity of semiconductor devices has driven the requirements of the microelectronic packaging technologies beyond that of simple space transformers. Increases in the number of interconnections, interconnection density, operating speed, driver switching, and power dissipation have forced the development of more sophisticated package technologies. This trend is particularly evident in the case of logic and microprocessor functions (also memory, to some extent), where package electrical and thermal performance have become critical.

The ability and extendibility of multilayer ceramic (MLC) chip carriers to meet these electrical and thermal performance requirements and compatibility with area array flip chip dies have contributed to the rapid rise of CBGA and CCGA packages. CBGA packages are competitive for some performance-driven applications with as few as 250 inputs-outputs (I/Os) and the package of choice for flip chips (FCs) with about 350 I/Os and beyond. CCGA packages provide one of the few options available in volume production for I/O requirements above 1000 and on a 1.27-mm pitch or less. CBGA and CCGA packages are ideal for satisfying high-interconnection-density requirements and a high-lead-count capability while being compatible with standard printed wiring board (PWB) ground rules and materials, surface-mount technology (SMT) card assembly, and well-established infrastructures.

9.1 CERAMIC BALL GRID ARRAY (CBGA)

CBGA packages are composed of four basic elements: a chip carrier or substrate, attached die, die or package encapsulation, and a solder ball interconnection structure. Each of these elements is described below. Typically, FC dies are used in CBGA packages. The key feature of the CBGA interconnection structure is the use of a 90Pb/10Sn solder ball (MP > 300°C) that does not melt during package or card assembly when joined with eutectic 63Sn/37Pb solder (Figs. 9.1 and 9.2). One of the larger body sizes is the 32.5-mm CBGA package with 625 solder balls on 1.27-mm pitch (Fig. 9.3).

9.1.1 Chip Carrier

Menu of Packages. The CBGA chip carrier typically consists of a standard MLC alumina substrate. The ceramic material poses no restrictions for package or card assembly processing because it is compatible with high temperatures and process chemicals used during package fabrication and assembly. Typically, the body sizes of CBGA packages range from 18.5 to 32.5 mm, corresponding to ball counts ranging from 196 to 625 on a 1.27-mm pitch and from 292 (depopulated) to 937 on 1.0-mm pitch. Depending on die complexity and package performance requirements, chip carriers typically range between 4 and 20 ceramic layers, corresponding to a fired substrate thickness of 0.8 and 3.0 mm, respectively.

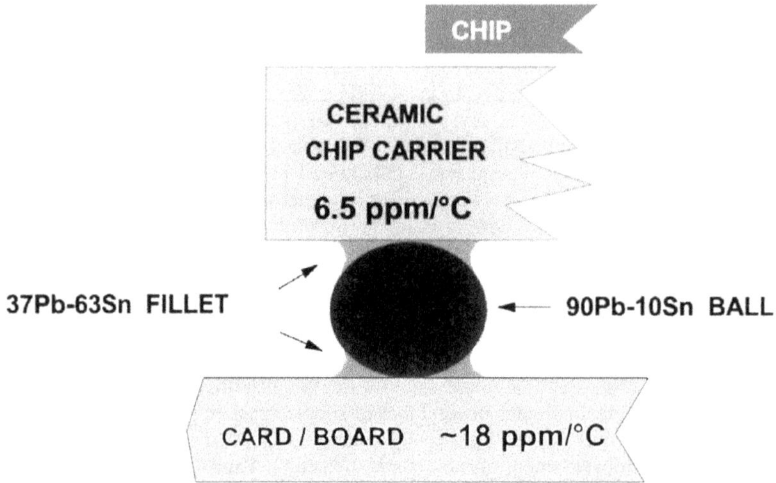

FIGURE 9.1 Schematic vertical cross section of a CBGA package joined to a PCB. (After T. Caulfield et al., Ref. [1], 1993.)

FIGURE 9.2 Metalloprahic vertical cross section of dual-metal solder joints attaching a CBGA package to PCB.

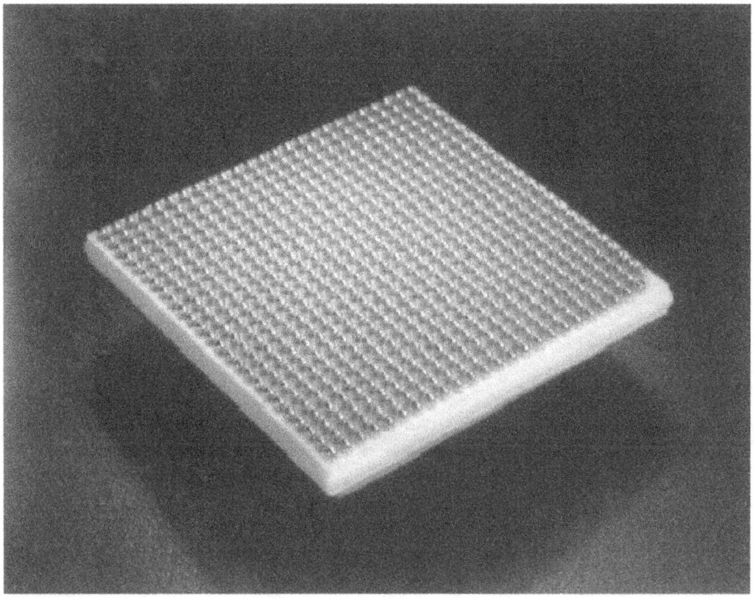

FIGURE 9.3 Photograph of a 32.5 mm CBGA package with 625 interconnections. (*Courtesy of IBM.*)

Ceramic Process. Individual ceramic layers or green sheets are punched with via patterns and screened with refractory metal–conducting paste to form lines and fill the vias. Depending on chip layout complexity, two or more layers are used to redistribute signals from chip connections through the package to the balls located on the bottom of the package. Power and ground planes are added as required to support voltage levels on the die and provide shielding between signal layers.

The personalization of individual ceramic layers is dictated by the die interconnection layout and package interconnection requirements. Typically, alumina chip carrier personalization consists of 0.08- to 0.10-mm-wide lines on a 0.15- to 0.20-mm pitch with less than 0.10-mm-diameter vias on a 0.23-mm pitch that provide the vertical communication path to attached dies and between ceramic layers.

The individual ceramic layers are laminated together and sintered at high temperatures to form a monolithic chip carrier. The last process step is overplating the die-attach pads and joining pads for solder ball attachment with nickel and gold. A generalized process flow for ceramic chip carrier fabrication starts with the casting of the raw material and ends with plating (Fig. 9.4). Personalization layers of the CBGA chip carrier typically can be categorized as signal redistribution or voltage planes (Fig. 9.5). A thorough description of the ceramic chip carrier and its processing can be found elsewhere in the literature.[3]

Materials Properties. Standard 92% alumina ceramic has a coefficient of thermal expansion (CTE) (6.5 ppm/°C) that is fairly well matched to silicon (3 ppm/°C). Solder interconnections between die and chip carrier, therefore, exhibit generally acceptable thermal cycle fatigue life, particularly with the use of die underfill materials that enhance fatigue life by an order of magnitude. The alumina ceramic CTE, however, is not so well matched to epoxy-glass PWB materials, such as FR4 (18 to 20 ppm/°C). This large CTE mismatch poses a reliability challenge for chip carrier to PWB interconnections. To achieve acceptable fatigue life, solder-joint structures joining ceramic chip carriers to PWBs must overcome significant strain. Fatigue life in area array solder joints is improved by optimizing geometry and materials factors such as joint height and alloy selection, respectively.

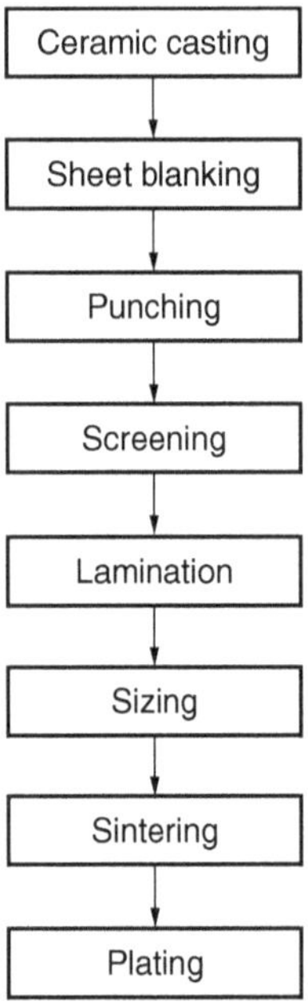

FIGURE 9.4 General process flow for ceramic chip-carrier fabrication.

CBGA and CCGA chip carriers are fabricated with either dark ceramic or white alumina ceramic. Newer ceramic materials also have been investigated recently, such as those with a higher CTE[4] or lower dielectric constant.[5]

9.1.2 Die Interconnection and Encapsulation

FC Dies. The FC proves to be the most advantageous die interconnection method for CBGA packages in terms of both performance and interconnection density. The MLC structure of chip carriers discussed earlier supports wiring-dense interconnection arrays by redistributing large numbers of signal connections and providing multiple power and ground connections. To achieve an FC attachment, a set of vias intercepts the surface MLC chip carriers and terminates at plated capture pads (termed *microsockets*) that match the array of solder bumps on an FC die. During placement, the die is faced downward, and the solder bumps are oriented to mate with corresponding chip carrier pads, thus the term *flip chip* (FC). Distortion and camber control of ceramic chip carriers across a die site are key factors in achieving high yields and reliable joints. During FC attachment reflow, 97Pb/3Sn die solder bumps are fully melted. Until recently, rosin-based flux and solvent cleaning had been required because of the high reflow temperature of high-lead solders. Advances in flux chemistries and joining processes, however, have allowed the implementation of no-clean fluxes for dies with a small number of solder bumps.

Die Underfill. CBGA FC applications are most often nonhermetic, with an epoxy encapsulant filling the gap between the die and ceramic chip carrier. The epoxy underfill completely surrounds all FC solder joints and strongly bonds to both the device and ceramic chip carrier. Underfill materials enhance thermal cycle solder-joint fatigue life by 10 times or more while providing environmental protection. In the case of capless or lidless configurations, a heat sink can be attached directly to the back side of the FC after card assembly with thermally conductive adhesive or by mechanical means.

Encapsulation Options. An encapsulation option for FC CBGA packages is to place an aluminum-alloy lid over the die, sealing it to the chip carrier with an adhesive, and filling the gap between die and lid with a thermal compound dispensed in the lid. The purpose of the thermal compound is to reduce internal thermal resistance. Adhesively joined lids will not pass a fine-leak test (i.e., are nonhermetic) but must pass a gross-leak test in order to preserve the thermal characteristics of the thermal compound. A heat sink may be attached after card assembly to provide additional cooling capability.

A variation on the nonhermetic lid structure is a lid attached directly to the backside of an FC, known as *direct lid attach* (DLA).[6] A flat lid is joined directly to the backside of an FC using a conductive adhesive material that serves to secure the lid to the die while providing a thermal path.

The choice of FC encapsulation options (Fig. 9.6) available for CBGA packages provides flexibility for a wide variety of applications. The lidless or DLA options allow CBGA packages to comply with J-STD-020 level 1 moisture specification requirements.[7,8] Nonhermetic lidded CBGA packages with FC dies typically meet the requirements of the Joint Electron Device Engineering Council (JEDEC) level 2.[9]

9.1.3 Package Interconnection Structure

The standard CBGA package interconnection structure is formed in two steps. High-melting-point solder balls are joined to chip carrier terminal pads with Sn/Pb eutectic solder. CBGA components are then placed on and reflowed to epoxy-glass PWBs via standard SMT.[10–12] The 90Pb/10Sn solder ball (MP > 300°C) does not reflow during package or card assembly (see Figs. 9.1 and 9.2). This

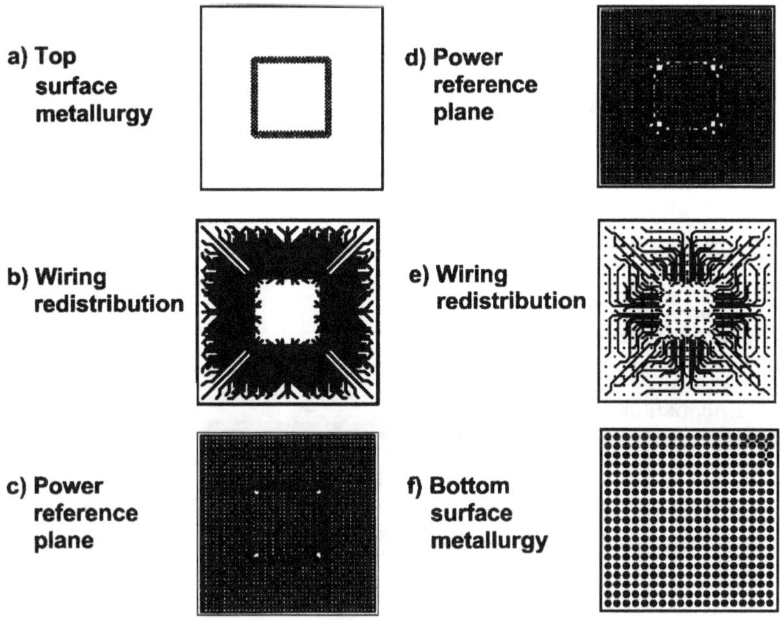

FIGURE 9.5 Various ceramic personalization layers for a CBGA chip carrier. g) Schematic vertical cross section depicting a laminated and sintered stack of ceramic layers into a monolithic chip carrier with an attached flip-chip die and package I/O solder balls.

hierarchical solder structure creates a consistent and reproducible gap or standoff between the ceramic package and the card. There is an inverse relationship between standoff height and strain generated during normal machine on/off cycles due to the thermal mismatch between the ceramic carrier and PWB. The thermal fatigue reliability of CBGA assemblies depends largely on the ability to accommodate and manage thermal-mechanical strains. Standoffs also provide clearance for cleaning after card assembly and help accommodate strains associated with card torquing and flexing during handling and box assembly. Use of meltable solder joints characteristic of plastic ball grid array (PBGA) packages that partially collapse during card assembly reflow would significantly decrease the fatigue life of CBGA packages joined to standard epoxy-glass PWBs. The increased joint height (i.e., standoff) of CBGA packages afforded by its high-melt structure aids in countering the effects of CTE mismatch between ceramic chip carriers (6.5 ppm/°C) and PWBs (18 to 21 ppm/°C) that would otherwise decrease fatigue life by two to three times.[13,14]

Typical CBGA packages use 0.89-mm-diameter balls on a 1.27-mm pitch. Increasing interconnection density by decreasing ball pitch requires smaller-diameter balls. For example, a 1.0-mm pitch requires the ball diameter to decrease to 0.8 mm. It is desirable to use the largest ball practical to obtain the highest standoff. Ball diameters must not be so large, however, that they create solder shorting concerns during package or card assembly. Metallized joining pads on chip carriers have a critical diameter. For example, 1.27-mm-pitch packages provide optimal manufacturability and reliability with approximately a 0.86-mm chip carrier pad diameter, whereas 1.0-mm-pitch packages require corresponding chip carrier pad diameters of 0.8 mm. The process for attaching solder balls to chip carriers is discussed in detail later in this chapter.

9.2 CERAMIC COLUMN GRID ARRAY (CCGA)

The CCGA package is a technology extension of the CBGA concept.[1,15,16] The solder ball array is replaced by an array of solder columns for improved thermal fatigue performance, required in the case of large ceramic body sizes. Solder columns (90Pb/10Sn) are joined with palladium-doped

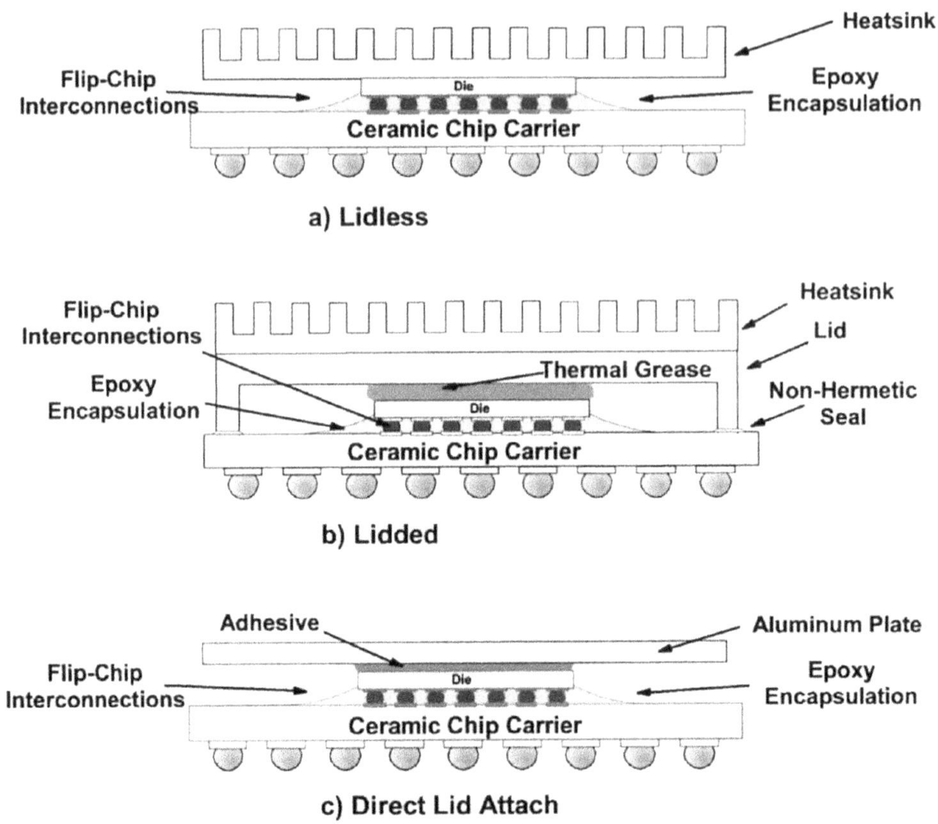

FIGURE 9.6 Schematic diagrams of various flip-chip CBGA packages. a) Lidless with a heat sink directly attached to the die after card assembly. b) Lidded with thermal compound to provide a thermal path between the die and lid-heatsink assembly. c) Lid directly attached to the die with an adhesive. (After J Zitz and F.L. Pompeo, Ref. [6], 1999.)

eutectic solder (Fig. 9.7). Similar to CBGA packages, the four basic elements of CCGA packages are the ceramic chip carrier, joined dies, encapsulation, and solder columns.

9.2.1 Chip Carrier

CCGA chip carriers are composed of the same multilayer alumina ceramic described earlier for CBGA packages. Because the primary purpose of a solder column is to extend solder-joint thermal fatigue life, CCGA body sizes typically start at 32.5 mm and currently extend to 45 mm, corresponding to 625 and 1224 I/Os a on 1.27-mm pitch, respectively. CBGA components in this range of body sizes would not be effective in dissipating strain at solder joints a large distance from the neutral point (DNP). The 42.5-mm CCGA package with 1088 I/Os (1.27-mm pitch) (Fig. 9.8), extendible to 1657 I/Os (1.0-mm pitch), illustrates the high I/O capacity of these packages. In addition, development is underway to further extend the CCGA package to a 52.5-mm body size providing more than 2500 I/Os. Since larger body sizes enable higher I/O counts, CCGA chip carriers are often more complex in terms of design (7 to 40 layers, 1.4 to 5.75 mm thick) than CBGA chip carriers.

9.2.2 Die Interconnection and Encapsulation

Similar to CBGA, the CCGA assembly processes of FC underfill and lid encapsulation are completed prior to column attachment. Encapsulation options for CCGA packages remain the same as for CBGA packages. Lids and/or heat sinks are more common with CCGA packages because the higher-function dies used with these packages dissipate more power.

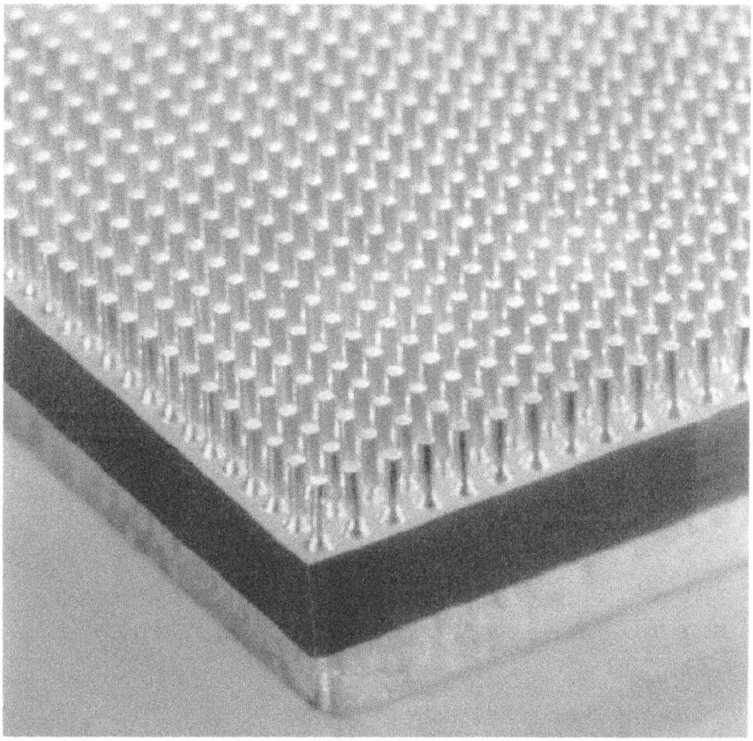

FIGURE 9.7 Legend to come.

9.2.3 Package Interconnection Structure

CCGA packages replace the high-melting-point solder ball of CBGA packages with a solder column of the same 90Pb/10Sn alloy. Solder columns, approximately 0.5 mm in diameter and either 2.2 or 1.27 mm tall, are joined to ceramic chip carriers with lead-doped Sn/Pb eutectic solder fillets in a manner similar to solder balls in the case of CBGA packages.

Because the column structure is less stable and more difficult to rework during card assembly than the ball structure, the lead is added to increase the robustness of the column interconnection in a process known as a *column last attach solder process* (CLASP).[16,17] The presence of lead in the eutectic solder creates a solder fillet with additional intermetallics that melt at a temperature higher than that of eutectic tin-lead (Fig. 9.9). These higher-melting intermetallics add structural support to the solder column joint during card assembly and rework processes.

After attachment, the columns are all uniformly shaved to the desired length on removal from the joining fixtures. With this exception, the CCGA process flow is identical to that of CBGA.

9.2.4 Effect of Pitch Reduction

For 1.0-mm-pitch packages, the only change in column structure and geometry developed for 1.27-mm-pitch CCGA packages is a reduction in the chip carrier pad diameter from 0.86 to 0.80 mm. Further reductions in column pitch, however, also would require an optimized geometry change to smaller-diameter columns because the 0.5-mm-diameter column used for 1.27- and 1.0-mm-pitch products would not easily be extendible to tighter pitches with chip carrier pad diameters smaller than 0.80 mm. It would be difficult to join columns to such small-diameter pads with consistency in a manufacturing environment because the columns are required to be captured within the perimeter of the pads.

FIGURE 9.8 Photograph of an IBM P2SC CCGA Package. (*Courtesy of IBM.*)

9.2.5 Thermal Cycle Fatigue Life

Tall, slender solder columns of CCGA packages are better able to accommodate strain arising from the CTE mismatch between a ceramic chip carrier and epoxy-glass PWB because the increased joint height increases the ease with which the solder joints can bend under shear conditions. Since joint strain is inversely proportional to joint height, the increased compliance of CCGA solder columns results in a significant enhancement in thermal cycle fatigue life over solder-ball joints. Thus, in addition to providing high I/O counts by virtue of the larger body sizes, solder-column packages also allow use in more severe application conditions where system temperature ranges or on/off cycles are more extreme.

9.3 PACKAGE STANDARDS AND OFFERINGS

9.3.1 CBGA JEDEC Standards

Two JEDEC outline standards, specified in the metric system, have been established for CBGA packages: MO-156 and MO-157, describing square and rectangular bodies, respectively. The features specified in the JEDEC outlines are package body size and tolerances, ball array size, total number of balls, ball diameter, coplanarity of the ball array, and true position (location tolerance) of the ball array. The overall package height is not specified because this dimension varies with chip carrier complexity and choice of encapsulation. The package size tolerance is ±0.2 mm, and the ball array coplanarity requirement is 0.15 mm. The location tolerance of the ball array is specified in terms of the geometric tolerance specification "true position."[18] The ball array true-position specification requires the center of each solder ball to be within a 0.30-mm-diameter circle (0.25 mm for 1.0 mm-pitch) whose center is defined by the design location relative to data defined by the package edges (Fig. 9.10). The ball array also has an offset tolerance, requiring the center of each ball to be

FIGURE 9.9 CLASP solder column joint. (*After S. Ray et al., Ref.* [17], 1999.)

within a 0.15-mm-diameter circle (0.10 mm for 1.0-mm pitch) whose center is defined by a best fit of the design ball array locations (see Fig. 9.10). The JEDEC specifications for ball coplanarity and true position have been evaluated to confirm that CBGA packages meeting these requirements can be assembled successfully.[19] Although not a JEDEC requirement, the A01 corner ball is often eliminated to easily identify the A01 corner.

The most common CBGA body sizes and their corresponding array sizes and I/O counts are given in Table 9.1. Body sizes range from 11 to 33 mm for square packages and 18.5 × 21 mm to 25 × 32.5 mm for rectangular packages. The available I/O range extends from 64 to 676 for CBGA packages with a 1.27-mm pitch. The JEDEC outline includes options for 1.5-, 1.27-, and 1.0-mm pitches, with the 1.27-mm pitch by far the most common. The I/O count for the largest CBGA package (33 mm) increases to 961 with a 31 × 31 array on a 1.0-mm pitch. Although the outline actually allows up to a 32 × 32 array for a 1.0-mm pitch on this 33-mm body size, processing and handling introduce concerns if balls are located too close to the edge of the ceramic carrier.

In addition, JEDEC standard MO-163, which defines the package outlines for PBGA memory components, is used for CBGA static random access memory (SRAM) packages. The SRAM package uses a 14-× 22-mm body size with either a 7 × 17 array (119 solder balls) or a 9 × 17 array (153 solder balls).

9.3.2 CCGA JEDEC Standards

Similarly, two JEDEC outline standards have been established for CCGA packages: MO-158 and MO-159, which describe square and rectangular bodies, respectively. The most common CCGA body sizes and their corresponding array sizes and I/O counts are given in Table 9.2. Body sizes range from 25 to 45 mm for square packages and from 25 × 32.5 to 32.5 × 42.5 mm for rectangular packages. The available I/O range extends from 361 to 1225 for a 1.27-mm pitch. While the 45-mm body size is included, with a few exceptions, the 42.5-mm square package is typically the largest CCGA in use. A 42.5-mm package (33 × 33 array) provides 1089 I/Os on 1.27-mm pitch, extendable to 1764 I/Os (42 × 42 array) if the pitch is reduced to 1.0 mm. To avoid processing and handling

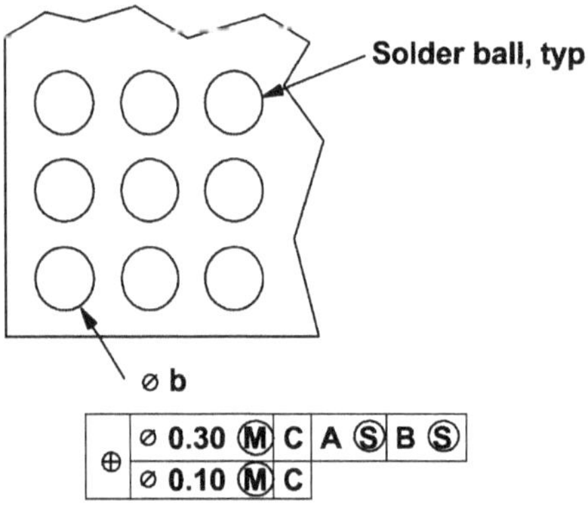

FIGURE 9.10 Ball true position as defined by JEDEC specification MO-156 and MO-157.

concerns, it is recommended that the 1.0-mm-pitch array on the 42.5-mm package be limited to 1657 I/Os (41 × 41 array with six columns removed in each corner).

9.4 TEST AND BURN-IN

Burn-in prior to ball or column attachment is recommended to prevent damage to nonmelting solder balls. This approach also minimizes value-added processes before burn-in. Many types of burn-in sockets are available to contact the package in this so-called land grid array (LGA) format. Final module test also can be performed prior to ball attachment if critical-speed sorts are not required. Otherwise, final module test can be performed after ball attachment using sockets with various contact schemes that minimize indenting or flattening of the solder balls. In some cases, burn-in may be performed after ball attachment, but it is critical that neither operation cause solder-ball or solder-column damage in violation of the JEDEC 0.15-mm coplanarity requirement or engineering specification.[20]

9.5 COMPARING CBGA AND CCGA

CBGA packages benefit from their low profile and the extremely robust nature of the ball array interconnection. For these reasons, CBGA packages tend to be chosen over CCGA packages. CBGA packages are limited, however, in terms of thermal fatigue reliability of the ball array interconnection. CBGA packages are reliable in body sizes up to and including 32.5 mm under typical desktop computer use conditions, considering combined effects of package temperature and number of system on/off cycles. CCGA packages are used beyond the 32.5-mm or smaller body sizes when CBGA packages fail to meet application requirements where either the use temperature range or the number of on/off cycles is excessive. CCGA packages may have either 1.27- or 2.2-mm-high columns to accommodate increased CTE mismatch–generated strains. Increases in interconnection height increase the thermal fatigue reliability of area array solder-joint interconnections.

TABLE 9.1 Array Size and I/O Count for Common CBGA Packages

	1.27-mm pitch		1.0-mm pitch*	
Body size (mm)	Array size	I/O count	Array size	I/O count
14.0 × 22.0	7 × 17	119	N/A	N/A
14.0 × 22.0	9 × 17	153	N/A	N/A
18.5 × 18.5	14 × 14	196	N/A	N/A
21.0 × 21.0	16 × 16	256	N/A	N/A
21.0 × 25.0	16 × 19	304	N/A	N/A
25.0 × 25.0	19 × 19	361	24 × 24	552
25.0 × 32.5	19 × 25	475	24 × 31	720
32.5 × 32.5	25 × 25	625	31 × 31	937

*Six balls are removed from the array in each corner.

While CCGA packages offer significant improvement in thermal fatigue reliability and are less sensitive to card assembly solder-volume requirements,[21] the columns create some difficulties. Handling concerns are increased because of column fragility. Columns also increase the package profile (height), an undesirable feature for products such as handheld electronics. For these reasons, CBGA remains the package of choice when reliability requirements permit.

9.6 COMPARING CBGA AND CCGA PACKAGES TO OTHER OPTIONS

9.6.1 Advantages

Interconnection Density: Full Array. An interconnection density advantage offered by CBGA or CCGA over most BGA packages is the ability to provide a full array of interconnections (i.e., balls, columns). Many BGA packages do not allow solder balls beneath the die area at the center of a package. Not only do CBGA and CCGA packages support the use of a full array, the multilayer structure allows a high degree of flexibility in assigning signal, power, or ground I/Os. These assignments can be made based on tradeoffs between electrical performance requirements and card wirability.

Thermal Performance. CBGA packages exhibit excellent thermal performance because of the inherently good thermal conductivity of ceramic chip carriers. The heat dissipation from BGA packages is divided among three paths: (1) through the chip carrier and solder balls into the PWB, (2) through the chip carrier into the cap and heatsink assembly, and (3) most directly through a path provided by a thermal compound between the die and heat sink into the airstream (Fig. 9.11*a*). The proportional heat flow among these paths depends on several factors, among them the heat-sinking capability of the card design (component density, number of metallized planes, and their proximity to the card surface). The heat sink design also affects the proportional heat flow. As the heat sink height increases, the percentage of heat flow into the card through the solder balls decreases (Fig. 9.11*b*).

In addition to the thermal paths through CBGA and CCGA chip carriers and their solder connections, FC packages can dissipate heat from the backside of the die through a thermally conductive compound to an aluminum lid or directly attached heat sink. In a typical desktop system, the CBGA package thermal resistance (Fig. 9.12) allows a 32.5-mm FC package to dissipate approximately 6 W without a heat sink, more than 12 W with a low-profile heat sink, and greater than 20 W with a large heat sink. Wire-bond CBGA packages can benefit thermally from a cavity-down configuration with or without a heat sink attached.

Electrical Performance. BGA packages, particularly in an FC format, provide short die-to-card signal paths, reducing electrical parasitics that affect signal time of flight and quality adversely. The importance of package electrical performance increases as the number of package interconnections increases.

Cofired molybdenum or tungsten personalization can provide strip-line signal conditions. Multiple power and ground planes provide impedance control for signal lines while reducing inductance along power paths.

TABLE 9.2 Array Size and I/O Count for Common CCGA Packages

	1.27-mm pitch		1.0-mm pitch*	
Body size (mm)	Array size	I/O count	Array size	I/O count
32.5 × 32.5	25 × 25	625	31 × 31	937
32.5 × 42.5	25 × 33	825	31 × 41	1247
42.5 × 42.5	33 × 33	1089	41 × 41	1657

Six columns are removed from the array in each corner

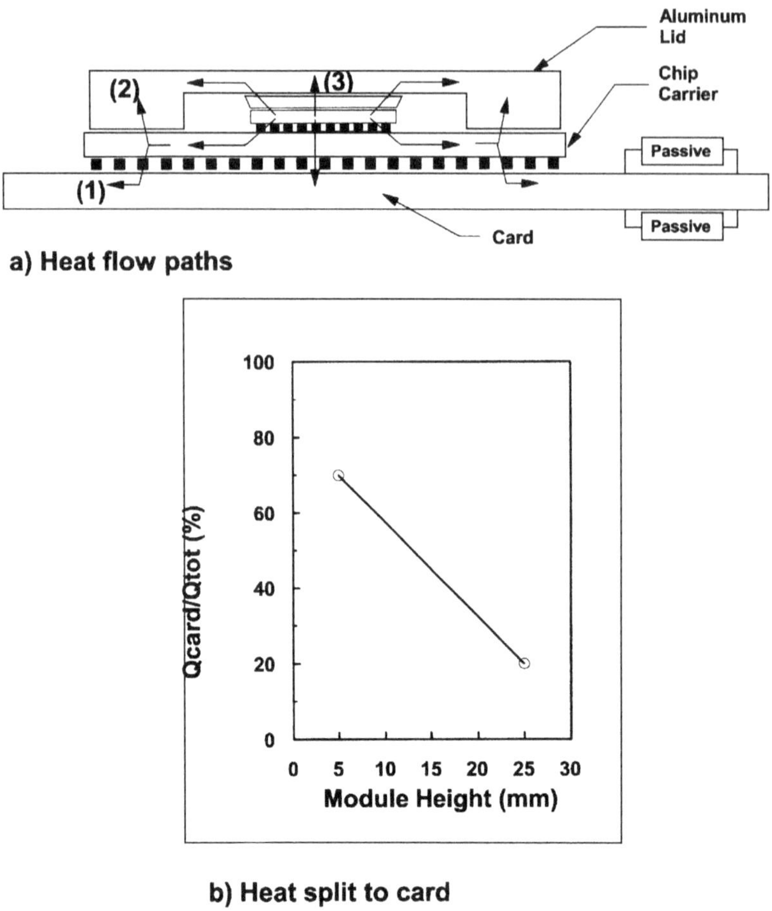

FIGURE 9.11 a) CBGA package heat flow paths. (b) Percentage of heat flow into card depending on heatsink height. (*Courtesy of J. Zitz, IBM Microelectronics.*)

CBGA packages are ideally suited to FC interconnections that reduce inductance by 1 to 2 nH over wire bonds. The multilayer structure also allows for wiring beneath the die, giving a full area array. These features provide CBGA packages with their high signal-to-power connection ratio, low signal-line noise, and high simultaneous switching capability. Modeling of CBGA package design ground rules indicates that clock frequencies into the gigahertz range can be supported. Ground rule improvements or new ceramic materials will allow frequencies higher than even 1 GHz to be achieved.

9.6.2 Disadvantages

The excellent I/O density and thermal and electrical performance attributes of CBGA and CCGA packages are all advantages in using these packages. Several disadvantages also must be considered. In the lower I/O or performance arenas, ceramic packages typically are not cost-competitive. At 256 I/Os or when voltage planes are required, the cost becomes competitive. While CBGA packages can be small (18 or 21 mm) and thin (chip carrier < 1 mm), the higher-performance CBGA and CCGA packages tend to be larger (32.5 or 42.5 mm) and thicker (chip carrier 2.5 to 4.0 mm). These large packages pose some challenges for the card assembly process in component handling and require reflow profile optimization. Although a full array capability provides an electrical advantage, it poses a wiring challenge in designing the PWB. And while the CTE of alumina chip

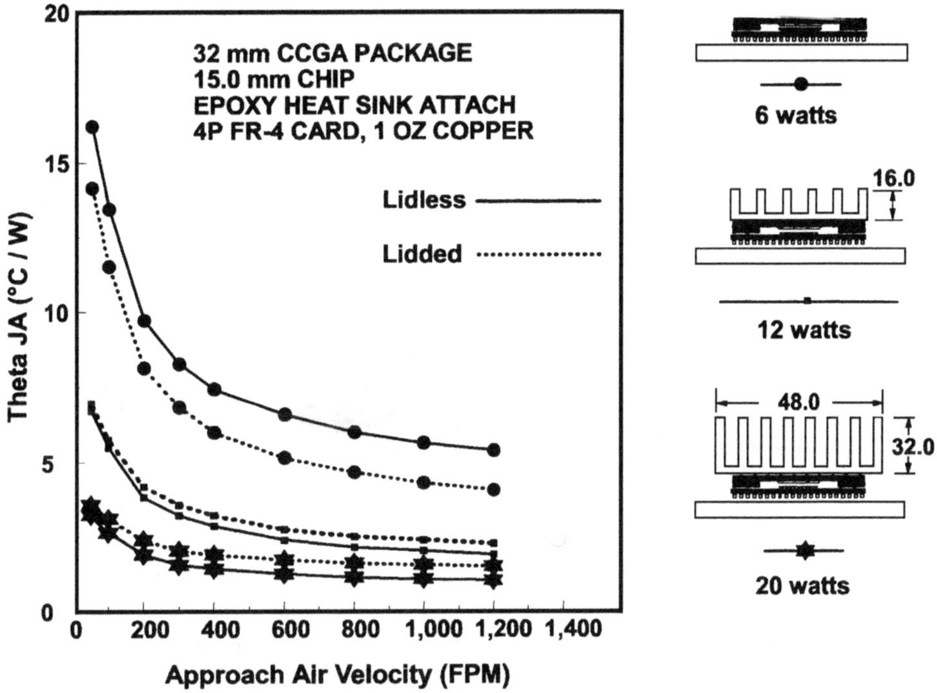

FIGURE 9.12 Thermal resistance for a 32.5 mm flip chip CBGA lidded and lidless packages for several heatsink and air flow conditions. (*After T. Caulfield et al., Ref.* [2], 1995. *Reprinted with permission of the McGraw-Hill Companies.*)

carriers supports highly reliable silicon die to package interconnections, it does limit the reliability of package to epoxy-glass PWB interconnections.

9.7 PWB REQUIREMENTS

As with other BGA packages, a dogbone card-attach configuration is required. There are key attributes and dimensions for designing dogbone card pads on 1.0- and 1.27-mm pitches for CBGA and CCGA packages (Fig. 9.13). Direct joining of BGA packages to card vias is not recommended. If solder runs into the via, depleting the card side fillet, it can result in a defective solder joint. There are techniques for maintaining a filled via to achieve direct joining, but none is currently practiced in production.

Design guidelines are found in several references.[22,23] The key elements of a dogbone card-pad design consist of a joining pad, plated through-hole (PTH) or via, wiring trace, and solder mask, as discussed below.

9.7.1 PWB Joining Pads

Joining pads are plated-copper surface pads to which BGA packages are attached by solder reflow. Solder paste is screened on these pads that are typically round but may be square or diamond-shaped.[24] The copper pads are formed as part of the personalization step during PWB fabrication with an etch angle causing the pad diameter at the bottom to be greater than the top by 0.025 mm (see Fig. 9.13, inset). While PWB fabricators tend to specify card-pad diameters at the pad bottom, PWB assemblers often refer to pad-top diameters. For reliable CBGA and CCGA solder joints on a 1.27-mm pitch, a minimum card-pad diameter of 0.68 mm (measured at the joining surface at the

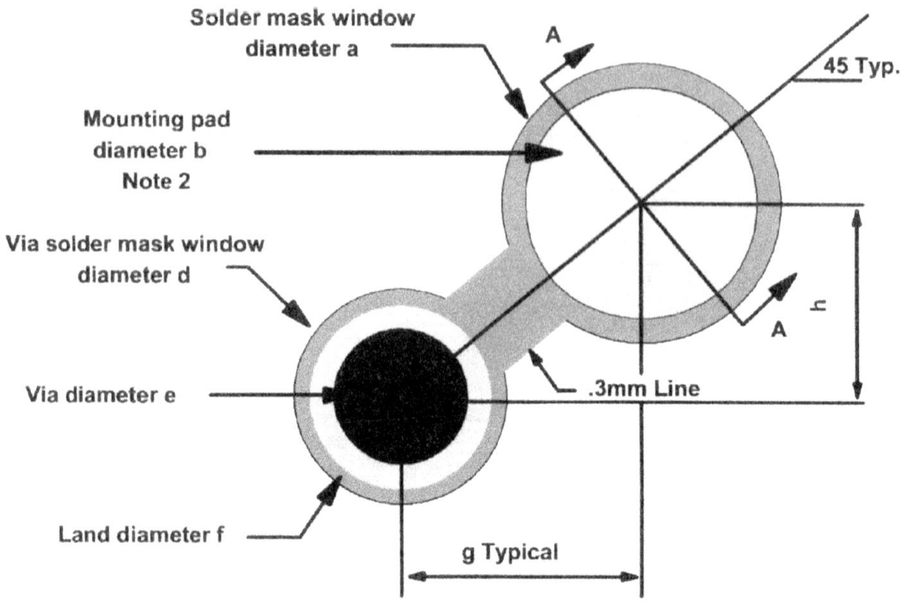

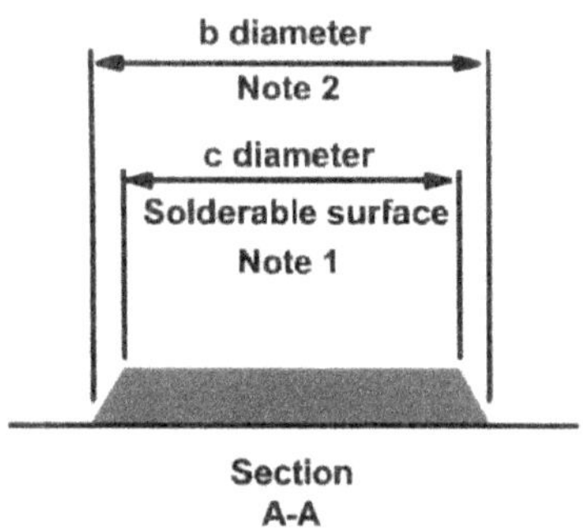

Feature	1.27mm Pitch	1.0mm Pitch
a	0.851 mm	0.80 mm
b	0.749 mm	0.69 mm
c	0.720 mm	0.67 mm
d	0.483 mm	0.38 mm
e	0.305 mm	0.20 mm
f	0.560 mm	0.46 mm
g	0.635 mm	0.50 mm
h	0.635 mm	0.50 mm

Note 1: Functional Surface
Note 2: Nominal Diameter at copper / FR4 interface with typica etch angle

FIGURE 9.13 Key attributes and dimensions of the card dogbone pad (non-solder mask defined, NSMD). (*After J. Stack, Ref.* [22], 1997.)

pad top) is required. The nominal card pad-top diameter is 0.72 mm, based on a typical tolerance range for card-pad diameters of ±0.04 mm. The maximum card-pad diameter is defined by the PWB fabricator's ground rule for copper-to-copper spacing. The closest allowable spacing typically is the distance between a joining pad and PTH land on a neighboring dogbone. Card pads larger than 0.72 mm, when allowed by supplier ground rules and capabilities, are desirable to further increase the reliability of CBGA and CCGA card side solder joints.

9.7.2 Plated Through-Hole (PTH)

Communication among inner layers of a card is made through vertically oriented via structures, called *plated through-holes* (PTHs), each connected to a surface pad by a copper wiring trace.

Typical PTH dimensions, to accommodate BGA/CGA components with a 1.27-mm I/O pitch, consist of 0.36-mm drilled holes plated to a 0.30-mm finished diameter. A copper pad with a diameter of 0.56 mm encircles the PTH at the card top and bottom surfaces. To maintain the largest viable card-pad diameter on a 1.0-mm pitch (0.69 mm nominally, measured at the bottom of the pad), the PTH size must be reduced to a 0.254-mm drilled hole and plated to a 0.20-mm finished diameter. In the case of very thick cards (>3.56 mm) for 1.27-mm-pitch arrays or even average-thickness cards (2.54 mm) for 1.0-mm-pitch arrays, the aspect ratio of PTH depth to diameter becomes a concern for PTH thermal fatigue life.[25] The quality of PTHs becomes critical in their ability to survive thermal exposures encountered during manufacture and assembly.

9.7.3 Wiring Trace/Solder Mask

Wiring traces connecting the joining pads to mating PTHs are typically 0.3 mm wide. The solder mask, to prevent solder flowing along the trace between the copper joining pad and PTH, must cover a distance of at least 0.08 to 0.10 mm between them to be effective. Care must be taken during card assembly rework not to damage the solder mask or underlying copper trace.

A solder mask is also used around joining pads. A solder mask opening larger than the copper joining-pad diameter (referred to as *non-solder-mask-defined,* NSMD; see Fig. 9.13) is preferred to an opening that is smaller than the copper surface pad (referred to as *solder-mask-defined,* SMD). The solder-joint appearance varies slightly between NSMD and SMD card pads (Fig. 9.14) because of this difference in solder mask configuration. A solder mask also may be used to "tent" or cover PTH openings to prevent fill during wave solder operations. When attempting to tent PTH vias, care must be taken to ensure that they are completely covered. Partial solder mask coverage can result in plugged vias if the solder mask material infiltrates the via, leading to reliability exposures of

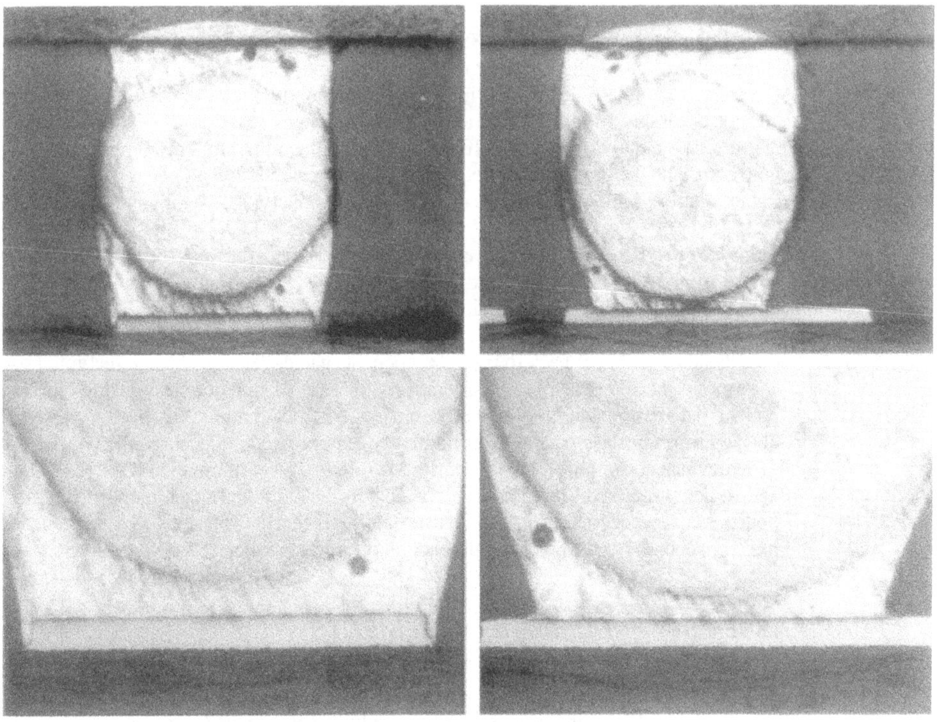

FIGURE 9.14 CBGA mounted on a non-solder mask defined (NSMD) pad (left) and a solder mask defined (SMD) pad (right). (*After R. Master et al., Ref.* [26], 1997. *Copyright 1997 IEEE.*)

weakened barrels from trapped chemical residues. Likewise, solder mask openings around PTHs must be large enough to prevent unintended plugging when the vias are designed to be open.

9.7.4 Card Surface Finishes

CBGA and CCGA packages are used with a variety of card surface finishes but most commonly with an organic solder protective (OSP) finish such as Entek. Other finishes include hot-air solder leveled (HASL) or nickel-gold (NiAu) over-plate layers. Any surface finish that is solder wettable and yields relatively flat pads is acceptable for reflow attaching of CBGA or CCGA components. Surface finish quality is important because it affects the wettability and thus reliability of an assembly.[27] The HASL process adds thermal exposures during raw card processing and, as noted earlier, is cause for concern when the card thickness to PTH diameter ratio is very high (>10). For example, PTH diameters of 0.30 mm in 3.56-mm-thick PWBs or PTH diameters of 0.20 mm in 2.54-mm-thick PWBs are considered challenging.

9.7.5 Ground Rules

Lines and Spaces. Card ground rules for 1.27-mm-pitch CBGA and CCGA packages are typically 0.127-mm (5 mil) lines and 0.127- (5 mil) or 0.152-mm (6 mil) spaces, allowing two wiring lines per channel on internal PWB layers. For higher-lead-count packages, such as 625 and above, three wiring lines per channel are achieved by choosing 0.076-mm lines (3 mil) and 0.10-mm (4 mil) spaces, which may be desirable to reduce the card layer count. For 1.0-mm-pitch packages, 0.076-mm (3 mil) lines and 0.102-mm (4 mil) spaces are necessary to achieve the required wiring density.

Layup/Flatness. The full array area format of CBGA and CCGA packages allows signal count and placement to be a tradeoff between card wirability and electrical performance, so card cross sections can range from four layers with two signal and two power planes (2S2P) to over 20 layers, with 12S10P as one example. The actual card cross section for any application depends on several factors, including number of signals, location of signals, number of voltage levels, card ground rules, component-to-component wiring requirements, and wiring efficiency. To reduce the chances of card warpage, a symmetric cross section is recommended. Overly restrictive card flatness requirements are not necessary to join CBGA and CCGA packages successfully. The industry standard range of 0.7 to 1.0 percent warpage (7 to 10 mils of warpage per inch of card length) is adequate as an incoming requirement because this specification is not intended to apply locally across a BGA site. The flatness across individual CBGA or CCGA card sites typically ranges from 0.025 to 0.10 mm.

Material/Layout. Any standard epoxy-glass PWB material may be used, where the CTE typically ranges from 16 to 22 ppm/°C, and where card thickness normally ranges from 1.37 to 3.56 mm. Card cross section and thickness, in addition to material factors, affect card CTE and stiffness, thus influencing the reliability of CBGA or CCGA interconnections. The layout format of CBGA packages (single- or double-sided, packages back to back, component spacing, etc.) also can affect interconnection reliability.

Among the PWB layout considerations for ease of manufacture during card assembly are component distribution and body-to-body spacing of CBGA and CCGA packages. Avoiding regions with clusters of high-mass-packages or relatively unpopulated areas eases the degree of tuning necessary to achieve an acceptable thermal profile for reflow. A minimum clear zone of 5.0 mm between package bodies is required for rework tooling clearance. Additional spacing up to 10.0 mm between package bodies may be necessary for large clusters of high-thermal-mass CCGA packages (e.g., 42.5 mm) to prevent secondary reflow of neighboring solder joints when reworking the CCGA packages.

9.8 CARD ASSEMBLY PROCESS REQUIREMENTS

The card assembly process and equipment for CBGA and CCGA packages reflect standard SMT requirements. Key card assembly steps include solder-paste screening, placement, solder reflow, cleaning, and inspection. Rework processes for the removal and replacement of CBGA and CCGA packages also have been established. Given that incoming packages that meet the solder-ball or solder-column coplanarity requirement of 0.15 mm and cards that meet the flatness requirements described earlier, establishing a high-yield card assembly process is generally straightforward. Many captive and contract assembly manufacturers have demonstrated high first-pass yields attaching CBGA components at card assembly compared with quad flat pack (QFP) components (Fig. 9.15). Process-control monitors or inspections at several steps ensure low defect levels (i.e., high yields) and high reliability. In particular, solder-paste screening must be monitored to ensure proper volume and quality of print, key variables in the process. The reflow profile also must be monitored to ensure a proper temperature profile at both center and outer solder joints. Card assembly process details for CBGA and CCGA packages can be found in industry literature.[21,28–30]

9.9 PACKAGE INTERCONNECTION RELIABILITY

9.9.1 Geometric Factors

CBGA. The geometry of attachment pads at both the ceramic chip carrier and PWB is the result of optimization studies that include experimentation and finite-element modeling.[31] The solder-ball and fillet compositions[14] and dimensions are also the result of extensive study.

Standoff Height. The solder-ball diameter is driven by a need to achieve a maximum standoff height to accommodate strains that result from the CTE mismatch between a ceramic chip carrier and epoxy-glass PWB. A high standoff height corresponds to a large ball diameter that is limited in size to allow consistent package and card assembly processing. A 0.89-mm ball diameter has been determined to be optimal for 1.27-mm-pitch components, whereas a ball diameter of 0.8 mm is optimal for 1.0-mm-pitch components.

Pad Diameter/Solder Volume. The chip carrier pad diameter needs to be sufficiently large to capture several electrical connections, or vias, within a ceramic carrier. On the other hand, the card-pad

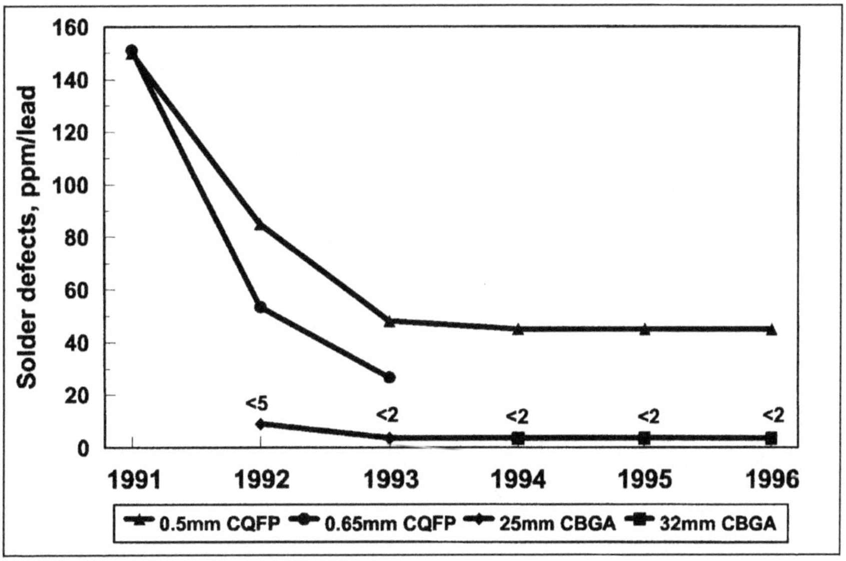

FIGURE 9.15 CBGA card assembly defect levels as compared to QFP. (*After T. Caulfield et al., Ref* [2], 1995. *Reprinted with permission of The Mcgraw-Hill Companies*).

diameter must be sufficiently small to meet PWB ground rules consistent with the PTH-dogbone structure and solder-mask requirements. Finite-element modeling (FEM)[31,32] and reliability testing[28] indicate a near-optimal structure with chip carrier and card pads at 0.86 and 0.72 mm, respectively, for a 1.27-mm pitch, with sufficient eutectic solder at both fillets. The recommended card-side solder-paste volume is 7000 cubic mils nominal, 4800 cubic mils minimum, for 1.27-mm pitch. The dependence of solder-joint reliability on both card-pad diameter and solder-paste volume is described by Phelan and Wang.[28] For card assembly of 1.0-mm-pitch CBGA packages, the recommended solder-paste volume is 2500 to 4600 cubic mils.[33]

CCGA. The card-pad diameter and solder volume again play a role in the interconnection reliability for both 1.27- and 1.0-mm-pitch interconnections,[21,34,35] with a recommended solder-paste volume of 3000 to 7600 and 2000 to 5000 cubic mils, respectively.

9.9.2 Metallurgical Factors

CBGA Joints. Metallurgical factors influence the reliability of CBGA solder joints. The choices of eutectic Sn/Pb solder fillets and 90Pb/10Sn solder balls are the result of metallurgical studies conducted to determine the optimal fatigue life within SMT processing constraints that require Sn/Pb-compatible alloys.[14] While silver and tin can be added to significantly increase the stiffness (i.e., creep resistance) of Pb/Sn eutectic solder fillets, it was determined that the fatigue life of dual-metal BGA solder joints is relatively insensitive to fillet stiffness. The choice of solder-ball alloy, however, has a very significant effect on joint fatigue life. The thermal fatigue life of joints with 97Pb/3Sn solder balls is about three times lower than that of joints with 90Pb/10Sn solder balls. The ductile, low-strength 97Pb/3Sn balls experience excessive deformation during thermal cycling, causing stress localization and distortion and resulting in early failure. CBGA joints consisting of 90Pb/10Sn balls and eutectic Sn/Pb fillets therefore provide an excellent balance: highly manufacturable, SMT compatible, and near-optimal fatigue life capability for alloys within the Pb/Sn system.

Single-Metal Solder Joints. Compared with an all-eutectic solder ball, a dual-metal system extends fatigue life by two to three times,[13,14] primarily due to the increased height of the noncollapsing dual-metal structure (37 versus 21 mils). The choice of single-metal alloy also influences the fatigue life.

CCGA Joints. Based on the aforementioned solder-ball alloy studies, 90Pb/10Sn solder is also used for solder columns. CCGA process studies optimized reliability,[21] including 1.0-mm-pitch components[34,35] and robustness of the solder-column interconnection to survive test and assembly.[36]

9.9.3 Additional Design Factors

More recent studies have investigated the influence of additional design factors, such as ceramic chip carrier thickness, that can have a significant effect on CBGA solder-joint reliability.[37] Based on modeling and experimental data, a 1.0-mm-thick chip carrier can have more than twice the fatigue life of a 3.0-mm-thick carrier, which is stiffer and less compliant. Factors such as the use of a lid, card thickness, and component spacing on the card, which also affect compliance, determine whether the full effect of chip carrier thickness is realized. Another aspect limiting thermal fatigue life is so-called green cycles arising from hardware or software power management adjustments to the operating temperatures and cycles, which was explored by Atwood et al.[38]

9.9.4 Mechanical Robustness

The effect of shock and vibration using MIL-810E–based conditions on both CBGA and CCGA solder-joint reliability has been studied to establish weight limits for adhesively attached heat sinks.[39] Weight limits for 32.5-mm, 625-I/O packages of 100 g for CBGA and 55 g for CCGA may be scaled based on the number of I/Os. Heat sinks attached with spring clips inserted into the PWB that apply a continuous compressive load to CBGA packages are common. In one study, a 44-g heat sink applying compressive loads of 1.8 to 5 kgf in conjunction with applied impact shocks of up to 20 and 75 g resulted in no degradation in fatigue life for a 25-mm CBGA package.[40]

9.10 SOLDER-BALL ATTACHMENT PROCESSES

9.10.1 Process Overview

The 90Pb/10Sn solder balls are joined to nickel- and gold-plated pads with 63Sn/37Pb eutectic solder. There are several process options both for eutectic solder application and for fixturing the solder balls. One approach to an automated ball attachment process flow shown in Fig. 9.16 is described in this section.[41]

Line automation requires integration of the various process sectors, in addition to prudent equipment and processing choices at each step. Belt conveyors link load stations, placement equipment, and reflow furnaces. Software controls the process flow at and between tools. High-speed vision-alignment systems can be used to place FC dies.[42]

In automating a CBGA ball attachment process to expand factory capacity, increase throughput, and decrease unit cost, the challenge is to maintain package quality. Key process and equipment choices must be made. The selection of either standard solder-paste printing or solder preform pretinning is one example.

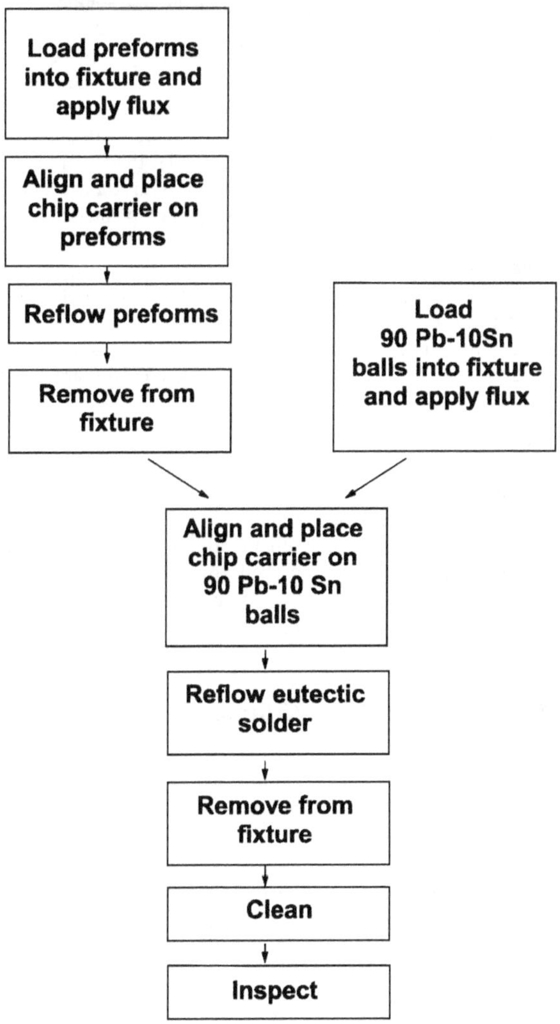

FIGURE 9.16 CBGA automated perform-process flow.

Screened Solder Paste. Concerns arise with paste screening, whether on ceramic chip carriers or on arrays of solder balls. If the choice is to screen on the ceramic chip carrier, dimensional concerns mentioned in the preceding section also apply here. To achieve alignment between chip carrier and balls, it is necessary to automate "flipping" the screened chip carriers. If the choice is to screen solder paste on a fixtured array of solder balls, then solder volume control continues to be a concern. To automate the process, screening parameters must be tightened and automated paste inspection implemented.

Solder Preform Process. A preform-based process changes the way Sn/Pb eutectic solder is supplied and alters the process flow of product through the ball attachment sector (see Fig. 9.16). Once optimized (i.e., adequately defining the sphere diameter), the preform process provides an extremely repeatable volume. Preforms are handled easily by automated equipment, allowing high production capacity.

Preform Loading. Preforms, which are small spheres of Sn/Pb eutectic solder, are loaded into universal (usable with a range of component body sizes) fixtures (Fig. 9.17), similar to the boats described previously. Large pallets are used with a preform fill pattern defined by form-factor stencils rather than dedicated fixtures. Following a controlled flux application, a standard SMT-type placement tool (with sufficient accuracy for fine dimensions) visually aligns the chip carrier to the array of preforms, placing fluxed chip carriers onto preform arrays. The assembly is placed in a reflow furnace to melt the eutectic solder preforms, pretinning the ceramic chip carrier's metallized pads.

Solder-Ball Loading. The 90Pb/10Sn solder balls are loaded into a universal pallet, fluxed, and await joining to pretinned chip carriers (see Fig. 9.17). Placement equipment, described earlier, is used to align pretinned chip carriers to a pallet loaded with solder balls. The flux must be sufficiently tacky to maintain the pretinned, slightly bump-shaped chip carrier pads centered on the solder balls.

Reflow Attachment. After aligning the solder balls to the mating chip carrier pads (prepared either with paste or preforms), the pallet assembly is placed in a reflow furnace as before, melting only the eutectic solder. Chip carriers are extracted from pallets and automatically loaded into cleaning fixtures. After cleaning, the CBGA packages are visually inspected, measured for dimensional specifications on a sampling basis, and then packed for shipment in either JEDEC trays or tape and reel format.

9.10.2 Solder-Ball Rework

Solder-ball arrays are reworkable if ball-related problems are discovered during fabrication or after removal from a PWB.[20] The ball rework process requires that all balls be removed through a squeegee reflow process (Fig. 9.18), leaving a thin solder layer on the chip carrier pads.[43] The operation is performed under a blanket of synthetic oil to prevent solder oxidation of the pads. After solder-ball and eutectic solder debris are removed from the chip carrier pads, a new array of solder balls is attached following the same process flow described earlier. Attempting to replace a single solder ball or use a chip carrier as removed from a card would risk using a ball array that does not meet coplanarity or true-position specifications.

9.11 SOLDER-COLUMN ATTACHMENT PROCESSES

CCGA interconnections are attached to ceramic chip carriers in a similar fashion to CBGA. The solder to form columns is supplied in the form of wire segments loaded into graphite boats rather than the solder ball.

9.11.1 CLASP Column Attachment Process for Automation

Automation of the CLASP column process is similar to that described for the CBGA process. The desire to maintain the CBGA process flow but increase the stability and robustness of the column to chip carrier solder joint during card assembly rework has led to the development of the column last attach solder process (CLASP).[16,17] The CLASP column attachment process replaces the standard eutectic tin-lead solder with a doped eutectic tin-lead solder. This doped eutectic solder creates a solder fillet with additional intermetallics that melt at a temperature higher than that of eutectic tin-lead

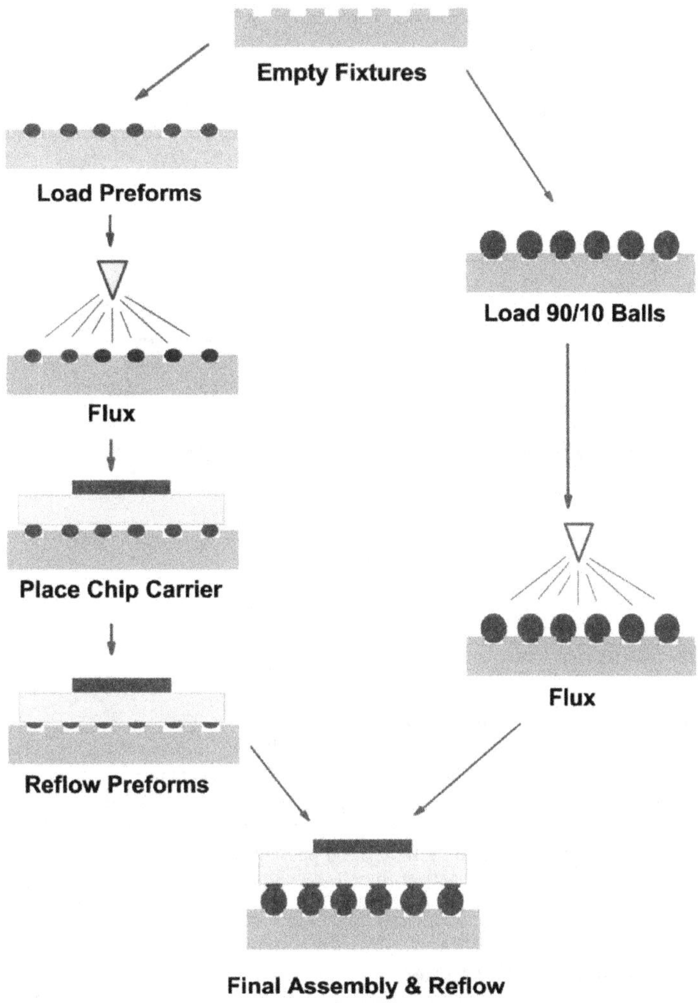

FIGURE 9.17 CBGA automated preform process schematic. (After J. Dankelman and L. Tousignant, Ref. [41], 1996.)

(see Fig. 9.9). These higher-melting intermetallics add structural support to the solder-column joint during card assembly and rework processes.

The CLASP process can be implemented in a highly automated manufacturing line. Solder-paste screening using an automated tool results in accurately located solder-paste deposits through vision alignment and high pad-to-pad solder volume uniformity, with 100 percent in-line paste inspection. The columns are also optimally centered on the chip carrier pads by the automated placement tool using vision and a best-fit algorithm. There is a significant reduction in handling of the in-process packages. An automated extraction and shave process with 100 percent automated inspection for column position after shaving also improves the quality of a finished solder-column array.

9.11.2 Column Rework

CLASP solder columns of CCGA packages can be reworked in a fashion similar to that described earlier for solder balls of CBGA packages.[20] The same rework process is used during package fabrication and after card removal.

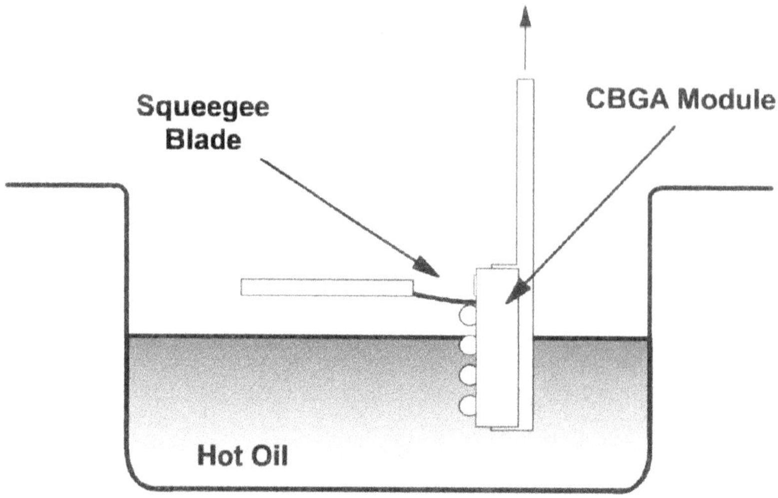

FIGURE 9.18 CBGA ball removal using a squeegee in a hot oil bath. (*After A. Downing et al., Ref* [43], 1997

9.12 SHIPPING CONTAINERS

9.12.1 CBGA Shipping Containers

Finished CBGA packages may be packed in molded JEDEC outline stacking trays compatible with automated card assembly placement equipment. The tray pockets are designed for a particular package size. Foam may be used to line the bottom of a tray, protecting the solder balls from damage. JEDEC trays have a diagonal cut at the A01 corner for ease of identification in loading tray stacks into card assembly placement equipment. CBGA packages are placed in JEDEC trays with their A01 corners in alignment with the tray A01 corner. Tape and reel packaging for high-volume applications is also an option for CBGA packages.

9.12.2 CCGA Shipping Containers

Protecting columns during package shipment and card processing is more challenging than for ball arrays. Care must be taken in loading and removing CCGA packages from trays. Ideally, a tray pocket suspends the columns above the pocket floor in an open space. If corner columns are removed or if an array size is reduced relative to the body size, the chip carrier can then rest on corner areas or edge ledges that are part of the tray design. Typically, I/O utilization required for the larger body sizes does not allow this luxury. For 1.27-mm grid packages, however, shipping tray pockets with a matrix of holes can be used if corner columns are necessary. An open pocket is required for 1.0-mm-pitch CCGA packages because mold-tolerance issues cause interference problems with a hole array. The best compromise allowing an effective tray design with the largest number of columns requires removing six columns in each corner. Corner areas rest on tray pocket ledges, which only modestly reduces the usable array size. Following these recommendations, 42.5-mm packages with 1.0-mm-pitch columns, limited to a 41 × 41 array with six columns removed in each corner, provide a total of 1657 columns.

9.13 FUTURE USE OF CBGA AND CCGA PACKAGES

Advances in CBGA and CCGA package technology are continuing to make these packages increasingly more attractive for a variety of applications. The trend toward increased package complexity will continue even at the low end. CBGA packages are as well suited to house commodities such as PC microprocessors and high-end products owing to the need for increased interconnection counts and density for products at all levels.

CCGA packages on 1.0-mm pitch and less will become more common to help fulfill future I/O requirements, although additional industry capability and cost parity for higher-density card technologies are necessary for their widespread use. High interconnection requirements for high-end servers and networking systems will be the key drivers.

An increased number of applications will require double-sided joining of CBGA and CCGA packages, including shared vias (at least for memory applications). These formats can result in lower solder-joint fatigue life and should have application-specific reliability testing and analysis.

9.14 SUMMARY

CBGA and CCGA packages are now standard in the microelectronics industry for cost-performance applications. The use of these packages in midrange servers, high-end workstations, and increasingly for network systems is driven by the need for high-I/O logic in high-I/O-density packages. CBGA and CCGA packages have demonstrated the ability to meet the electrical and thermal performance requirements of these applications while providing a highly manufacturable and reliable FC package. Growth in the use of these packages is expected to continue as new systems drive to even higher I/O and performance requirements. While other FC packages emerge in the marketplace, these ceramic packages will continue to meet the needs of advanced systems.

ACKNOWLEDGMENTS

This chapter would not have been written without the technical guidance and editorial encouragement of Karl Puttlitz. His support is especially appreciated. In addition, many IBM colleagues over the years have made significant contributions to the understanding and documentation of numerous topics relating to CBGA and CCGA packaging, including J. Acocella, D. Banks, R. Behun, J. Benenati, P. Brofman, A. Caron, T. Caulfield, J. Corbin, J. Currie, J. Dankelman, S. Dwyer, L. Goldmann, H. Harrington, L. Heck, C. Heim, K. Hoebener, T. Holmes, E. Ingalls, M. Interrante, P. Isaacs, R. Jackson, J. Jaspal, L. Jimarez, J. Jozwiak, E. Kastberg, S. Konecke, G. Martin, R. Master, C. Milkovich, G. Phelan, H. Quinones, J. Ross, S. Ray, C. Reynolds, W. Sablinski, J. Stack, K. Stalter, L. Tousignant, T. Wiggins, and W. Wildey. Special thanks to Tom Rednour for assistance with the figures.

9.15 REFERENCES

1. Caulfield, T., Benenati, J. A., and Acocella, J., "Surface Mount Array Interconnections for High I/O MCM-C to Card Assembly," in *Proceedings of the 1993 International Conference and Exhibition on Multichip Modules,* Denver, CO, April 1993, pp. 320–325.
2. Caulfield, T., Cole, M. S., Cappo, F., Zitz, J., and Benenati, J., "An Overview of Ceramic Ball and Column Grid Array Packaging," in J. H. Lau (ed.), *Ball Grid Array Technology,* McGraw-Hill, New York, 1995, Chap. 5.
3. Tummala, R. R., Garrou, P., Gupta, T., Kuramoto, N., Niwa, K., Shimada, Y., and Terasaa, M., "Ceramic Packaging," in R. R. Tummala, E. J. Rymaszewski, and A. G. Klopfenstein (eds.), *Microelectronic Packaging Handbook,* Chapman & Hall, New York, 1997, Chap. 9.
4. Yamaguchi, K., Higashi, M., Hamada, N., Yonekura, H., and Kunimatsu, Y., "Improvement of Solder Joint Reliability between Multilayer Ceramic Package and Printed Wiring Board by New Ceramic Material," in *Proceedings of the 47th Electronic Components and Technology Conference,* San Jose, CA, May 1997, pp. 1277–1282.
5. Fasano, B. V., Indyk, R., O'Connor, E., Plachy, A. L., and Reddy, S. N. S., "Glass Ceramic Substrates for Flip Chip Packages," *Future EMS International,* 1999, pp. 24–26.

6. Zitz, J., and Pompeo, F. L., "Direct Lid Attach Packaging for Ceramic Chip Carrier Applications," *IBM Microelectronics MicroNews,* 5 (2), pp. 24–26, 1999.
7. "J-STD-020, Moisture/Reflow Sensitivity Classification for Plastic Integrated Circuit Surface Mount Devices," EIA / JEDEC JC-14.1, Electronic Industries Association, Englewood, CO, October 1996.
8. "EIA / JEDEC Standard Preconditioning of Nonhermetic Surface Mount Devices Prior to Reliability Testing," JESD22-A113-B, Electronic Industries Association, Englewood, CO, March 1999.
9. Coffin, J., "Moisture Sensitivity of CBGA and CCGA Packages," in *Proceedings of International Electronics Packaging Society Conference,* Austin, TX, September 1996, pp. 94–101.
10. Behun, J. R., Cole, M. S., Hoebener, K. G., Call, A. J., Cappo, F. F., Milliken, J. C., and Klingel, B. T., "Improved Interconnection Structure and Test Method," U.S. Patent No. 5,060,844, 1991.
11. Behun, J. R., Cole, M. S., Hoebener, K. G., Call, A. J., Cappo, F. F., Milliken, J. C., and Klingel, B. T., "Interconnection Structure and Test Method," U.S. Patent No. 5147,084, 1992.
12. Cole, M. S., and Caulfield, T., "Ceramic Ball Grid Array Packaging," *Advancing Microelectronics,* ISHM, Jan.-Feb. 1995, pp. 18–19, 52–56.
13. Banks, D. R., Burnette, T. E., Gerke, R. D., Mammo, E., and Mattay, S., "Reliability Comparison of Two Metallurgies for Ceramic Ball Grid Array," in *Proceedings of the 1994 International Conference on Multichip Modules,* Denver, CO, April 1994, pp. 529–534.
14. Puttlitz, K. J., Caulfield, T., and Cole, M. S., "Effect of Material Properties on the Fatigue Life of Dual Solder (DS) Ceramic Ball Grid Array (CBGA) Solder Joints," in *Proceedings of the 45th Electronic Components and Technology Conference,* Las Vegas, NV, May 1995, pp. 1005–1010.
15. Banks, D. R., Heim, C. G., Lewis, R. H., Caron, A., and Cole, M. S., "Second-Level Assembly of Column Grid Array Packages," in *Proceedings of SMI,* San Jose, CA, August 1993, pp. 92–98.
16. Farooq, S., Interrante, M., Ray, S., and Sablinski, W., "Interconnect Structure and Process for Component Assembly and Rework," SMTA Surface Mount Technology Assoc., Docket No. 9-97-180, filed 1/98.
17. Ray, S. K., et al., "CLASP Ceramic Column Grid Array Technology for Flip Chip Carriers," in *Proceedings of the 2nd Annual Packaging Symposium at Semicon West,* San Jose, CA, July 1999, pp. F1–F7.
18. *Dimensioning and Tolerancing,* Y14.5M, ASME, New York, 1999.
19. Cole, M. S., and Hoebener, K. G., "Planarity and Centrality Requirements for Ceramic Ball Grid Array Packaging," in *Proceedings of Surface Mount International,* San Jose, CA, August 1995, pp. 273–278.
20. *CBGA and CCGA Module Limitation/Inspection Specification,* IBM Microelectronics, Hopewell Jct., NY, November 2000.
21. Phelan, G., Welch, M., Wang, S., and Cole, M., "Card Assembly and Reliability of 44-mm Ceramic Solder Column Array Modules," in *Proceedings of Nepcon West,* Anaheim, CA, February 1995, pp. 1048–1058.
22. Stack, J., *BGA/CGA Raw Card and Assembly Design Guide Covering CBGA, CCGA, TBGA, and PBGA,* IBM Document No. EN01-2711-05, March 1997.
23. Johnston, P., "Printed Circuit Board Design Guidelines for Ball Grid Array Packages," in *Proceedings of Surface Mount International,* San Jose, CA, August 1995, pp. 255–260.
24. Yee, S., and Ladhar, H., "The Influence of Pad Geometry on Ceramic Ball Grid Array Solder Joint Reliability," in *IEEE/CPMT International Manufacturing Technology Symposium,* Austin, TX, October 1996, pp. 267–273.
25. Knadle, K., and Ferrill, M. G., "Failure of Thick Board Plated Through Vias with Multiple Assembly Cycles: The Hidden BGA Reliability Threat," in *Proceedings of Surface Mount International,* San Jose, CA, September 1997, pp. 109–115.
26. Master, R. N., Dolbear, T. P., Cole, M. S., and Martin, G. B., "Ceramic Ball Grid Array for AMD K6 Microprocessor Application," in *Proceedings of the 48th Electronic Components and Technology Conference,* sponsored by SMTA Surface Mount Technology Assoc., Seattle, WA, May 1998, pp. 702–706.
27. Bradley, E., and Banerji, K., "Effect of PWB Finish on the Reliability and Wettability of Ball Grid Array Packages," *IEEE Transactions on Components, Packaging, and Manufacturing Technology, Part B,* 19 (2), May 1996, pp. 320–330.
28. Phelan, G., and Wang, S., "Solder Ball Connection Reliability Model and Critical Parameter Optimization," in *Proceedings of the 43rd Electronic Components and Technology Conference,* Orlando, FL, June 1993, pp. 858–862.
29. Heck, L., Lewis, R., and Phelan, G., "Surface Mount Assembly and Rework for Ball Grid Array Packages," in *Proceedings of New and Critical Technologies for Surface Mount,* Raleigh, NC, October 1994, pp. 11–14.
30. Heck, L., "Card Assembly Rework for Ceramic Column Grid Array Packages," in *Proceedings of Nepcon West,* Anaheim, CA, February 1995, pp. 1971–1975.
31. Corbin, J. S., "Finite Element Analysis for Solder Ball Connect (SBC) Structural Design Optimization," *IBM Journal of Research and Development,* 37 (5), September 1993, pp. 585–596.
32. Guo, Y., and Corbin, J. S., "Reliability of Ceramic Ball Grid Array Assembly," in J. H. Lau (ed.), *Ball Grid Array Technology,* McGraw-Hill, New York, 1995, Chap. 8.

33. Cole, M., Duchesne, R., Interrante, M., and Jimarez, L., "Design and Process Optimization for 1.0 mm Pitch CBGA," in *Proceedings of SMTA International,* San Jose, CA, September 1999.
34. Cole, M., and Berger, K., Brofman, P., "Design and Process Optimization for 1.0 mm Pitch CCGA," in *Proceedings of SMI,* San Jose, CA, August 1998, pp. 7–13.
35. Ingalls, E. M., Cole, M. S., Jozwiak, J., Milkovich, C., and Stack, J., "Improvement in Reliability with CCGA Column Density Increase to 1 mm Pitch," in *Proceedings of the 48th Electronic IEEE Components and Technology Conference,* Seattle, WA, May 1998, pp. 1298–1304.
36. Ingalls, E. M., and Jozwiak, J., "Robustness of Deformed Ceramic Column Grid Array Columns," in *Proceedings of IMAPS,* Philadelphia, PA, October 1997, pp. 623–627.
37. Martin, G. B., Cole, M. S., Brofman, P. J., and Goldmann, L. S., "The Effect of Substrate Thickness on CBGA Fatigue Life," in *Proceedings of Surface Mount International,* San Jose, CA, September 1997, pp. 172–177.
38. Atwood, E., Collins, E., and Quinones, H., "Area Array Interconnect Reliability Prediction Using Stochastically Defined Conditions," in *International Symposium on Microelectronics,* Minneapolis, MN, October 1996, pp. 538–543.
39. Cole, M. S., Kastberg, E. J., and Martin, G. B., "Shock and Vibration Limits for CBGA and CCGA," in *Proceedings of Surface Mount International,* San Jose, CA, September 1996, pp. 89–94.
40. Dolbear, T. P., Master, R. N., Cole, M. S., and Martin, G. B., "Effect of Mechanical Shock and Vibration on the Second-Level Temperature Cycling Reliability of Ceramic Ball Grid Arrays with a Continuous Compressive Load Applied," in *Proceedings of IMAPS,* Philadelphia, PA, October 1997, pp. 385–389.
41. Dankelman, J., and Tousignant, L., "Automation at 1st Level Assembly of Ceramic Ball Grid Array Packages," in *Proceedings of the 2nd International Assembly and Packaging Foundry Conference,* Sunnyvale, CA, 1996.
42. DiSciullo, J., and Glidden, W., "High Volume Production of Ceramic BGA Packages," *Electronic Packaging and Production,* 37 (12), September 1997, pp. 91–94.
43. Downing, A. J., Foster, D. C., and Puttlitz, K. J., "Apparatus and Method for Removing Meltable Material from a Substrate," U.S. Patent No. 5,620,132, 1997.

INDEX

ABOUT THE EDITOR

KEN GILLEO, PH.D., is a well-known author/editor/columnist in the fields of circuitry, MEMS, and microelectronics packaging, and CEO of ET-Trends LLC, a consultancy and intellectual property company focused on emerging technology. His *Area Array Packaging Handbook: Manufacturing and Assembly*, also published by McGraw-Hill, is the premier professional guide in the field. Dr. Gilleo also has edited the *Handbook of Flexible Circuits* and *Polymer Thick Film*, and has written over 400 articles on packaging, circuitry, MEMS, and materials. He holds 35 U.S. patents, currently serves on the Board of Directors of the Surface Mount Technology Association, and is active in IEEE, IMAPS, and other professional organizations. He lives in Warwick, Rhode Island.

www.ingramcontent.com/pod-product-compliance
Lightning Source LLC
LaVergne TN
LVHW080652011026
849160LV00014B/1399